Interferometry - Recent Developments and Contemporary Applications

Edited by Mithun Bhowmick and Bruno Ullrich

Published in London, United Kingdom

IntechOpen

Supporting open minds since 2005

Interferometry - Recent Developments and Contemporary Applications
http://dx.doi.org/10.5772/intechopen.74004
Edited by Mithun Bhowmick and Bruno Ullrich

Contributors
Naiting Gu, Chandrasekhar Roychoudhuri, Vladimir (or Uladzimir) P Minkovich, Joel Villatoro, Pavel Minkovich, Ramon Jose Perez, Alexander P. Vladimirov, Mithun Bhowmick, Bruno Ullrich

First published in London, United Kingdom, 2019 by IntechOpen
IntechOpen is the global imprint of INTECHOPEN LIMITED, registered in England and Wales, registration number: 11086078, The Shard, 25th floor, 32 London Bridge Street
London, SE19SG - United Kingdom
Printed in Croatia

British Library Cataloguing-in-Publication Data
A catalogue record for this book is available from the British Library

Additional hard and PDF copies can be obtained from orders@intechopen.com

Interferometry - Recent Developments and Contemporary Applications
Edited by Mithun Bhowmick and Bruno Ullrich
p. cm.
Print ISBN 978-1-78985-926-3
Online ISBN 978-1-83880-297-4
eBook (PDF) ISBN 978-1-83880-298-1

For EU product safety concerns:
IN TECH d.o.o., Prolaz Marije Krucifikse Kozulić 3, 51000 Rijeka, Croatia, info@intechopen.com or visit our website at intechopen.com.

Meet the editors

Dr. Mithun Bhowmick obtained his PhD in Physics from Virginia Tech and subsequently worked at the Nazarbayev University and the University of Illinois, Urbana-Champaign. In 2019 he joined the Department of Mathematical and Physical Sciences, Miami University. His primary research areas are spectroscopy under dynamic shock compression, optoelectronic properties of semiconductors, and strength of materials. Dr. Bhowmick has over 20 publications.

Dr. Bruno Ullrich graduated from University of Vienna with a PhD in Physics and served on the Physics faculty at Bowling Green State University, Ohio (USA) and at at National University of Mexico for many years as professor of physics. His current research interest is thermo-optical properties of II-VI and III-V materials. He is an expert in micro- and nanofabrication techniques, as well. Dr. Ullrich has over 180 peer-reviewed publications.

Contents

Preface

The aim of this book is to provide a broad range of topics on interferometry. The topics include some modern techniques, recent approaches to solve issues with a technique, applicability of interferometry in novel scenarios, as well as analytical treatments needed to pursue in-depth exploration.

In recent times, the application of interferometry has broadened. With that in mind, an overview of updates and recent contributions is provided in Chapter 1, the introductory chapter. It is intended to highlight the significance of recent works as well, with emphasis on connections between other scientific areas of research that have been key to the relevance of the technique. In Chapter 2, radial shearing interferometer is discussed. Chapter 3 describes a detailed method to differentiate superposition principle from measurable superposition effects. Fiber-optic ring resonator is discussed in Chapter 4. Chapter 5 highlights modal interferometers based on a tapered special photonics crystal fiber for highly sensitive detection. Finally, in Chapter 6, application of interferometry in biological systems is addressed through dynamic speckle interferometry technique.

It was an exciting journey for us to work on this project. We would like to thank all the contributors for sharing their expertise on the topic and for their commitment to academic integrity. We want to acknowledge our author service manager, Ms. Sara Debeuc, and other staff of IntechOpen for their untiring support and patience, which made the book possible.

Finally, we would like to thank the readers; we sincerely hope this book achieves its goals by contributing positively to the research community.

Mithun Bhowmick
School of Chemical Sciences,
University of Illinois,
Urbana, USA

Department of Mathematical and Physical Sciences
Miami University,
Ohio, USA

Bruno Ullrich
Ullrich Photonics LLC,
Ohio, USA

Chapter 1

Introductory Chapter: Interferometry

Mithun Bhowmick and Bruno Ullrich

1. Introduction

Interferometry has been a very effective tool for science and industry for many years. From optics to astronomy, from materials science to mechanical engineering, interferometry has been instrumental in pushing the boundaries of technological progress. The field has come a long way since one of its earliest mention in the news section of *Nature* in 1920 [1]. At one time optical interferometry relied heavily on on monochromators, limiting the potential applications. Nowadays, the electromagnetic source used in generating fringes is far advanced, owing to the advent of lasers with unprecedented bandwidths and intensities. However, the techniques associated to interference of electromagnetic waves is no longer limited to all optical methods. Incoherent sources can now be made to interfere under certain experimental conditions [2].

The power of interferometry has made it an obvious tool to explore interdisciplinary research in modern times and has expanded to topics related to chemistry, mechanical engineering and material science related topics such as shock spectroscopy, surface profiling, and microfluidics [3–5]. In the recent gravitational wave discovery, it was an interferometer called "LIGO" that made it possible. In this chapter, a brief discussion will be presented on how the field has evolved in different research directions with time along with some examples from the recent literature which has helped the expansion. A brief overview of issues of interferometry technique in general will also be provided where artifacts inherent with the technique will be discussed.

Interferometry represents a set of techniques where, in its most common form, electromagnetic waves superimpose thereby forming interference fringes. The phenomenon involving interference occurs all around us. For example, a drop of oil on water or a soap bubble generates coloration because of naturally occurring interference. In interference, small changes in optical path cause significant and precisely measurable changes in intensity pattern. Using this property, a wide variety of experiments is designed where precise length measurements are necessary. In mechanochemistry, shock science, and detonation studies, measuring free surface velocity is very important. Optical interferometry with the highest time resolution has enabled such measurements. These techniques fall under the umbrella of "laser velocity interferometry" and consist of well-known and established methods such as VISAR, ORVIS, several Quartz-based stress gauges and so on [6]. A similar technique, termed as photon Doppler velocimetry (PDV) has been very useful to probe explosive chemistry in the short length scale with nanosecond time resolution [3]. These interferometric techniques can also accurately investigate shock response in solids. In this technique, typically a single mode laser beam is split into two. One of them is focused onto a moving, flat surface. The surface is usually polished; hence a strong reflection of the beam is generated. The reflected beam combines with the remaining beam, which has gone through a delay line meanwhile, to generate beats. The beats are counted using suitable algorithm to make a velocity history of the surface (i.e., the reflector). The same experiment can be performed in two different geometries, the most common being one where we can record a velocity history for an incoming object. In shocked systems, one can thus obtain information regarding pressure and density evolutions with high time resolution [3, 6]. In another geometry, a thin layer of reflective surface (typically a metal) is deposited at one side of the substrate being shocked, and the substrate is exposed to shock in such a way that the shock emerges through the reflective surface. This is a very powerful technique to generate shock break out profiles and is also used in generating Hugoniot points in materials. A Hugoniot is a collection of data points obtained from single shot shock experiments. Each data point in a Hugoniot curve gives us a shock velocity (U_s) corresponding to a mass particle velocity (U_p). Using this information, one can distinguish between shock loading and unloading mechanisms in solids. In a recent study, the researchers have shown that shock break out experiments in explosives are significantly affected on solid properties if the solids are different in crystallinity [7]. Thus, a highly crystalline sample such as sapphire or calcium fluoride would show different shock break out traces compared to polymer or glassy substances [7]. Gauge experiments are also reputed for accurately generating Hugoniot points. A series of such experiments performed LASL data center is currently being used by the shock community for impedance matching in designing experiments [8].

Interdisciplinary research has seen a great deal of hybrid techniques where two powerful spectroscopic methods are simultaneously used to extract unique information from a system. Fluorescence interferometry (FI) is one such example. There are various configurations of FI. In one variant, the self-reference interferometer consists of lenses that gather light from two sides of the sample simultaneously after the main beam is split using suitable optics [9]. The two beams are used as excitation and probe [9]. In another version, a Fourier transformed signal is used to generate a time domain information [9]. In general, fluorescence interferometry is very powerful as it complements the traditional fluorescence experiments by extracting the phase information. Such experiments have already provided with high resolution single shot images in biological specimens with marking capability [9]. Several ongoing and future research efforts are focused on getting a reliable, detailed 3D image using selective markers with the help of quantum dots. In summary, more advanced techniques such as optical nanoscopy is where FI is going to be very useful.

Another established area of research that benefitted greatly is materials science laboratories where optical materials are characterized using either white-light-interferometry or where interferometry is used for surface profiling [10, 11]. Both two- and three-dimensional optical profilers use interferometry to accurately map surface roughness. In the most common set up, an optical profiler uses the optical path difference to generate interference fringes between two signals coming from a reference surface and a "signal" surface. The "signal" surface is basically the surface under investigation. Many commercial optical profilers (for example, Zygo) are in operation in modern characterization laboratories [11]. Surface profiling is very crucial for understanding stress models in solids. There is a huge demand of modeling solids under stress. A lot of efforts are being made currently in those directions, where one would be able to numerically simulate response of solids [12]. To do that, the model should be robust enough to successfully simulate existing experimental data. There are open questions related to chemical kinetics and initiation of energetic materials which could be answered once the models can accurately simulate current high-resolution interferometry data [13]. One such example is the reactive and flow model "ALE3D," developed at Lawrence Livermore National Laboratory [14].

In fields such as industrial sensing and space engineering, measuring distances with high precision is important, where a traditional form of single-wavelength interferometry has been used in the past [15]. To get the distance, a fringe counting algorithm is used in single-wavelength interferometry, which sometimes can cause errors in position measurements [16]. However, a relatively new technique called multiple-wavelength interferometry has been in practice recently which avoids fringe counting [17]. The technique has lots of advantage such as capability of measuring very small distances, flexibility, etc., with one small disadvantage; it requires several synthetic wavelengths for increasing the experiment range of the system, thereby increasing complexity of the technique. To solve this problem, researchers have started to use a frequency comb, which emits equally spaced ultra-short pulses [17]. These pulses are spread over a broad spectrum with discrete and uniform modes. The modes are also spaced equally, with very narrow lines [17]. This resembles an ultra-accurate ruler, with deviation in the order of 70 nm for a measurement of 100 mm length [19]. Frequency combs are generated in different ways such as 4-wave mixing or using a mode locked laser, or even by electro-optic modulation. In 2005, Theodor W. Hänsch and John L. Hall shared half of the Nobel prize in Physics for the discovery of a group of precision spectroscopy techniques, including optical frequency comb [20].

There are several applications of precision interferometric methods in metrology, materials science, satellite technology, and a myriad of similar fields where optoelectronic and photonic properties are utilized. As mentioned earlier while briefly discussing white light interferometry (WLI), a series of novel quantum optical techniques employed for reducing noise in a superposition signal, to avoid third-order dispersion, and thrive on interferometric equalization [21]. The resulting variant of WLI is called quantum-WLI (or Q-WLI). Q-WLI is a fast, all-optical technique for investigating material properties with high precision [10].

Despite its versatility and impact, there are some areas where optical interferometric effects make spectroscopy very difficult to perform. One such example is a coherent artifact occurs in ultrafast degenerate pump-probe experiment with femtosecond pulses, where the strong signal at temporal overlap of the two beams screens the actual rise time of the "true signal" [21]. A degenerate pump-probe experiment (DPP) is the case where both the pump and probe are of same wavelengths [22]. DPPs are heavily used in probing novel materials to extract signatures of many-body effects in electron/spin scattering mechanisms [21, 22]. By varying the relative intensities of two beams, or by adjusting the polarization angle that effect can be corrected [23, 24]. Interference patterns generated by unwanted resonance can thus affect optical measurements. There are various experimental scenarios of light-matter interaction where signal to noise ratio needs to be very high. Raman scattering measurements, photoluminescence and fluorescence experiments, quantum confinement studies, spin injection and manipulation studies, absorption spectroscopy, and many such complex and delicate probing techniques rely on efficient collection of photons. It is important to understand the limitations in those collection processes as the measurement is designed, since interference of signals can obscure the data one is looking for. For example, while measuring transmission spectra from a thin film, or a layered material, interference effects are observed as oscillations in the spectra [25]. Sometimes, the effect is welcome when the film thickness is unknown and needs to be measured [26, 27]. At other times the interference effect makes it difficult to efficiently fit the spectra to get important band parameters in solids [28]. In photoluminescence measurements, such effects can change the linewidth and the analysis based on broadening of the signal gets affected [29, 30]. In micro and nano-fabrication techniques, and various laser based thin-film deposition systems, interference effects are also visible [30]. A recent paper attempted to analyze intrinsic photoluminescence stoke shift in pulsed laser deposited thin film of CdS to interpret physical meaning of a phenomenological parameter used in Urbach model [30]. The same paper also tried to look at the Raman peaks and probed the phonon population by calculating Huang-Rhys factor [30]. Raman lines are particularly susceptible of background noise, and there are many techniques developed for enhancing Raman signal. A recent study concluded that interference effects are crucial in the micro-Raman spectroscopy of graphene and must be included in analysis when extracting various material information from the spectra [31]. As we move into ternary and quaternary alloys and magnetically doped novel materials for ambient operable devices that successfully exploit charge and spin mobility, experimenters have come to avoid any such effects in the measurement scheme [32].

On February 11, 2016, observation of gravitational waves was announced from the Advanced Laser Interferometer Gravitational-Wave Observatory (LIGO) [33]. It has become one of the most significant scientific landmarks of recent times, prompting a Nobel prize in 2017 to Rainer Weiss, Kip Thorne and Barry Barish for their contribution towards this discovery [34, 35]. A detailed description of LIGO is out of the scope of this chapter. In short, the apparatus is a Michelson interferometer built at a massive scale [35]. Just like the Michelson-Morley experiment made a conclusive discovery in 1887 to disprove "Luminiferous Aether," LIGO-interferometer measurements have conclusively detected presence of a gravitation waves. It is clear from the updates presented above, how interferometry is one of the most advanced and precision techniques presently being pursued. The development in the field of laser technology has made the progress rapid. Very high time resolutions are being increasingly attained using attosecond laser pulses [36]. This is another huge improvement in understanding reaction mechanisms and will enable scientists to understand processes involved in solar cells by watching in real time the detailed step-by-step energy flow/conversion [36].

To summarize, interferometry is undoubtedly one of the most thriving and rapidly progressing experimental techniques. With applications in so many different fields, interferometers are going to be employed with increasing fashion. The versatility makes the technique almost indispensable in high precision spectroscopy and novel materials. The small and large-scale applicability made it an automatic choice as a consistent and transferable tool. The fact that two of the historic discoveries were made in the field of Physics on interferometer vouches for its power and significance. With increasing finesse in laser engineering, ultra-short laser-pulse based interferometers are going to resolve fundamental mechanisms of physical and life sciences with even greater detail, and this will impact a few of the most urgent technological needs modern society face today: renewable energy/energy harvesting, incorporating devices in biomaterials, and scarcity of resources.

Author details

Mithun Bhowmick[1,3]* and Bruno Ullrich[2]

1 School of Chemical Sciences, University of Illinois, Urbana, USA

2 Ullrich Photonics LLC, Wayne, Ohio, USA

3 Department of Mathematical and Physical Sciences, Miami University, Ohio, USA

*Address all correspondence to: bhowmick@illinois.edu

References

[1] Applications of interferometry. Nature. 1920;**104**:677. Available from: https://www.nature.com/articles/104677b0

[2] Patel R, Achamfuo-Yeboah S, Light R, Clark M. Widefield two laser interferometry. Optics Express. 2014;**22**:27094-27101. DOI: 10.1364/OE.22.027094

[3] Bhowmick M, Nissen EJ, Dlott DD. Detonation on a tabletop: Nitromethane with high time and space resolution. Journal of Applied Physics. 2018;**124**:075901. DOI: 10.1063/1.5043540

[4] Chen LC, Nguyen XL. Measurement science and technology. 2010;**21**(5)

[5] Domachuk P, Grillet C, Ta'eed V, Mägi E, Bolger J, Eggleton BJ, et al. Microfluidic interferometer. Applied Physics Letters. 2005;**86**:024103. DOI: 10.1063/1.1849415

[6] Forbes JW. Shock Wave Compression of Condensed Matter: A Primer. New York: Springer; 2012. DOI: 10.1007/978-3-642-32535-9

[7] Bhowmick M, Basset WP, Matveev S, Salvati L III, Dlott DD. Optical windows as high-speed shock wave detectors. AIP Advances. 2018;**8**:125123. DOI: 10.1063/1.5055676

[8] Marsh SP, editor. LASL Shock Hugoniot Data. University of California Press; 1980

[9] Bilenca A, Cao J, Colice M, Ozcan A, Bouma B, Raftery L, et al. Fluorescence interferometry: Principles and applications in biology. Annals of the New York Academy of Sciences. 2008;**1130**:68-77. DOI: 10.1196/annals.1430.038

[10] Kaiser F et al. Quantum enhancement of accuracy and precision in optical interferometry. Light: Science & Applications. 2018;**7**:17163. DOI: 10.1038/lsa.2017.163

[11] https://www.zygo.com/?/met/profilers/surfaceprofiling.htm

[12] Stekovic S, Nissen EJ, Bhowmick M, Stewart DS, Dlott DD. Numerical predictions of shock propagation through unreactive and reactive liquids with experimental validation. AIP Conference Proceedings. 2018;**1979**(1):100039. DOI: 10.1063/1.5044911

[13] Bhowmick M, Nissen EJ, Matveev SM, Dlott DD. Studies in shocked nitromethane through high dynamic range spectroscopy. AIP Conference Proceedings. 2018;**1979**(1):100004. DOI: 10.1063/1.5044876

[14] Nichols AL III et al. ALE3D v4.14 Manual Volume 2, Material and Chemical models. Livermore, CA: Lawrence Livermore National Laboratory; 2011

[15] Bender PL, Hall JL, Ye J, Klipstein WM. Satellite-satellite laser links for future gravity missions. Space Science Reviews. 2003;**108**:377-384. DOI: 10.1023/A:1026195913558

[16] Falaggis K, Towers DP, Towers CE. Multiwavelength interferometry: Extended range metrology. Optics Letters. 2009;**34**:950-952. DOI: 10.1364/OL.34.000950

[17] Dandliker R, Salvade Y, Zimmermann E. Distance measurement by multiple-wavelength interferometry. Journal of Optics-Nouvelle Revue D Optique. 1998;**29**:105-114. DOI: 10.1088/0150-536X/29/3/002

[18] Kim SW. Combs rule. Nature Photonics. 2009;**3**:313-314. DOI: 10.1038/nphoton.2009.86

[19] Wu G, Liao L, Xiong S, Li G, Zhu Z. Synthetic wavelength interferometry of an optical frequency comb for absolute distance measurement. Scientific Reports. 2018;**8**:4362. DOI: 10.1038/s41598-018-22838-0DO

[20] https://en.wikipedia.org/wiki/Frequency_comb#Applications

[21] Bhowmick M, Khodaparast GA, Mishima TD, Santos MB, Saha D, Sanders G, et al. Interband and intraband relaxation dynamics in InSb based quantum wells. Journal of Applied Physics. 2016;**120**(23):235702. DOI: 10.1063/1.4971347

[22] Bhowmick M, Merritt TR, Khodaparast GA, Wessels BW, McGill SA, Saha D, et al. Time-resolved differential transmission in MOVPE-grown ferromagnetic InMnAs. Physical Review B. 2012;**85**(12):125313. DOI: 10.1103/PhysRevB.85.125313

[23] Hui L, Hang Z, Jin-Hai S, Li-He Y, Fengand C, Xun H. Elimination of the coherent effect in a pump-probe experiment by directly detecting the background-free diffraction signal. Chinese Physics Letters. 2011;**28**(8). DOI: 10.1088/0256-307X/28/8/086602

[24] Lebedev MV, Misochko OV, Dekorsy T, Georgiev N. On the nature of "coherent artifact". Journal of Experimental and Theoretical Physics. 2005;**100**(2):272-282. DOI: 10.1134/1.1884668

[25] Heavens OS. Optical Properties of Thin Solid Films. New York: Dover Publications; 1954

[26] Ullrich B, Antillón A, Bhowmick M, Wang JS, Xi H. Atomic transition region at the crossover between quantum dots to molecules. Physica Scripta. 2014;**89**(2):025801. DOI: 10.1088/0031-8949/89/02/025801

[27] Ullrich B, Singh AK, Barik P, Xi H, Bhowmick M. Inherent photoluminescence Stokes shift in GaAs. Optics Letters. 2015;**40**(11):2580-2583. DOI: 10.1364/OL.40.002580

[28] Ullrich B, Bhowmick M, Xi H. Relation between Debye temperature and energy band gap of semiconductors. AIP Advances. 2017;7(4):045109. DOI: 10.1063/1.4980142

[29] Ullrich B, Singh AK, Bhowmick M, Barik P, Ariza-Flores D, Xi H, et al. Photoluminescence lineshape of ZnO. AIP Advances. 2014;**4**(12):123001. DOI: 10.1063/1.4897383

[30] Ullrich B, Ariza-Flores D, Bhowmick M. Intrinsic photoluminescence Stokes shift in semiconductors demonstrated by thin-film CdS formed with pulsed-laser deposition. Thin Solid Films. 2014;**558**:24-26. DOI: 10.1016/j.tsf.2014.02.047

[31] Yoon D, Moon H, Soon Y, Choi JS, Park BH, Cha YH, et al. Interference effect on Raman spectrum of graphene SiO_2/Si. Physical Review B. 2009;**80**:125422. DOI: 10.1103/PhysRevB.80.125422

[32] Meeker MA, Magill BA, Merritt TR, Bhowmick M, McCutcheon K. Dynamics of photoexcited carriers and spins in InAsP ternary alloys. Applied Physics Letters. 2013;**102**(22):222102. DOI: 10.1063/1.4808346

[33] Castelvecchi D, Witze A. Einstein's Gravitational Waves Found at Last. Nature/News [Internet]. February 11, 2016. Available from: https://www.nature.com/news/einstein-s-gravitational-waves-found-at-last-1.19361 [Accessed: January 12, 2019]

[34] Overbye D. Gravitational Waves Detected, Confirming Einstein's Theory. New York Times [Internet]. February

11, 2016. Available from: https://www.nytimes.com/2017/10/03/science/nobel-prize-physics.html [Accessed: January 12, 2019]

[35] LIGO's Interferometer [Internet]. Available from: https://www.ligo.caltech.edu/page/ligos-ifo [Accessed: January 12, 2019]

[36] Schnabi J. The World's Shortest Laser Pulse. Physics.org [Internet]. October 27, 2017. Available from: https://phys.org/news/2017-10-world-shortest-laser-pulse.html [Accessed: January 12, 2019]

Chapter 2

Radial Shearing Interferometer

Naiting Gu and Qun Luo

Abstract

Radial shearing interferometer (RSI) is one of the most powerful tools in many domains, especially in optical testing. RSI has compact size and good vibration immunity, which is adaptive to various environments, due to its common-path configuration. Moreover, it is very convenient application because no plane referencing wavefront is needed. The disadvantages of the conventional RSIs are that the distorted wavefront is hard to extract quickly and accurately from one radial shearography due to the phase extract algorithm is complex. Fortunately, the new RSIs can receive benefits from the accuracy of the methods of phase-shifting interferometry, and phase-shifting shearography is more sensitive than simple digital shearography. There are two mainly trend to the RSIs based on phase-shifting technique, i.e. instantaneous phase-shifting and compact size. In this chapter, a development process of RSI will be introduced briefly firstly, and then the some new RSIs based phase-shifting techniques in our work will be described in following parts, including initial RSI by using four-step polarization phase-shifting, modal wavefront reconstruction method for RSI with lateral shear and a new kind of compact RSI based micro-optics technique.

Keywords: radial shearing interferometer, phase-shifting technique, instantaneous, vibration insensitive, modal wavefront reconstruction

1. Introduction

Radial shearing interferometer (RSI) was proposed firstly in 1961 [1]. After development of many years, the RSI has been used widely in optical testing [2–4], corneal topographic inspection [5–7], wavefront sensing [8–11] and laser beam characterization [12–14]. Radial shear can be introduced by some classical optical components [15, 16] and the other different ways [17–21] including optical gratings [17, 18], zone plate [19], speckle interference [20, 21] and the other applications. It is very convenient application because no plane referencing wavefront is needed [24], especially comparing to the point diffraction interferometer (PDI) [22, 23]. RSI has compact size and good vibration immunity, which is adaptive to various environments, due to its common-path configuration. Recently, RSI has been becoming one of the most important tools for diagnosing the wavefront of laser beams and the other applications. The disadvantages of the conventional RSIs are that the distorted wavefront is hard to extract from only one radial shearography due to the complexity of the phase extract algorithm. However, fortunately, many of these applications can receive benefits from the high accuracy of phase-shifting interference methods, and phase-shifting shearography is more sensitive than simple radial shearography [25]. Thus some authors [8, 26–28] have proposed several kinds of RSIs by using temporal or spatial phase-shifting techniques. A cyclic RSI was developed in Ref. [26] for phase shifting interferometry with a polarization phase shifter. A RSI based on a Mach-Zehnder configuration by using liquid-crystal-device as the phase shifter was described in Ref. [27]. A Sagnac RSI with geometric phase-shifting technique was designed in Ref. [28]. However, all above these developed RSIs adopted the temporal phase shifting technique. In this case, the wavefront under test must be essentially stationary during the duration of acquiring several phase-stepping images, and they cannot be used to measure dynamic wavefront. In Ref. [8], a new RSI with spatial phase shifting techniques was developed for wavefront measurements and diagnosing dynamic measurements and the single-shot common-path phase-stepping configuration improved the performance of RSI especially for dynamic scenes. However, its complex configuration and extraction algorithm limits the testing accuracy, system stability, compactness and its application. Apparently, it is highly desirable to acquire compact one-shot vibration-insensitive RSIs. In Ref. [29], a new compact RSI with simultaneous phase-shifter based on a micro-retarder array (MRA) was proposed, which has some good features such as high speed, high accuracy and vibration immunity while decreasing the complexity. In Ref. [11], a new modal reconstruction algorithm was proposed to reconstruct accurately and quickly the wavefront under test in a RSI with or without lateral shear.

In this chapter, we will introduce the new RSIs with instantaneous phase-shifting and its corresponding modal wavefront reconstruction method. These works have been published earlier in [8, 11, 29], and they are re-arranged in this manuscript. This manuscript is organized as follows. In Section 2 a single-shot common-path phase-stepping RSI will be introduced. A new RSI with simultaneous phase-shifter based on a micro-retarder array will be described in Section 2.1. In Section 3 a modal wavefront reconstruction method based on Zernike polynomials in a RSI with or without lateral shear will be described. The conclusion of this chapter will be made in final section.

2. RSI based on simultaneous polarization phase-shifter

RSIs in many domains can receive advantages from the accuracy of phase-shifting interference methods, and many new RSIs were proposed by utilizing temporal or spatial phase-shifting techniques. Comparatively, the temporal phase shifting technique requires stationary scene during the duration of acquiring several phase-stepping images, but the spatial phase-shifting technique not. Thus the RSIs with spatial phase-shifting technique are preferred to apply in several applications due to its ability for dynamic scene. In this part we will introduce a new RSI based on simultaneous polarization phase-shifter.

2.1 System design and basic principle

Figure 1 shows the schematic diagram and its experimental setup of the RSI based on simultaneous polarization phase-shifter. It is composed of three main individual parts, i.e. wavefront simulator (WS), Hartmann-Shack wavefront sensor (HS WFS) and common-path phase-shifting RSI, which is the core part of this system. The proposed RSI includes a four-channel polarization phase stepper (FCPPS) and a cyclic RSI, which comprise a polarizing beam splitter (PBS1), two lenses (L5 and L6) and two flat mirrors (M1 and M2). The cyclic RSI can be treated as a Keplerian telescope system comprised L5 and L6. P2 is used to filter the state of polarization (SoP) of incident light at a fixed filtering angle. The passing beam will be guided into the RSI system and divided into two beams according to its

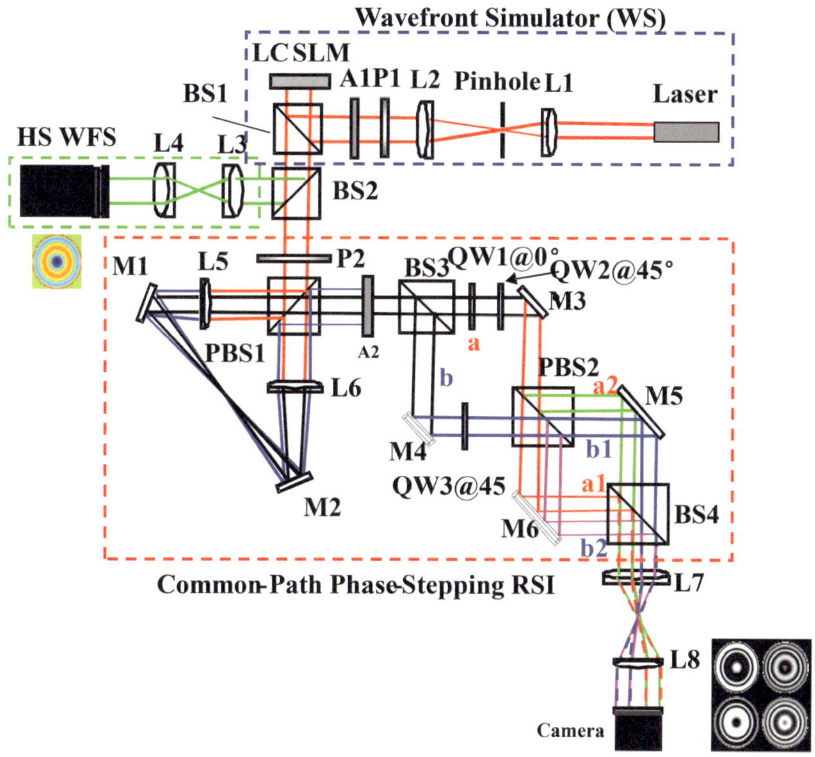

Figure 1.
Concept design and experimental setup of the RSI with spatial phase-shifting method. L1–8, lenses; PBS1/2, polarization beam splitters; BS1–4, intensity beam splitters; M1–6, mirrors; QW1–3, quarter-wave plates; A1–2, diaphragms; P1–2, polarizers; WFS, wavefront sensor.

components of SoP. According to the property of PBS, the reflected and transmitted lights has vertical linear SoP and horizontal linear SoP respectively, they will transmit the cyclic RSI system along anticlockwise direction (i.e. PBS1 → L5 → M1 → M2 → L6 → PBS1) and clockwise direction (i.e. PBS1 → L6 → M2 → M1 → L5 → PBS1) respectively and are guided into FCCPS system. These two beams are magnified and de-magnified due to the different focal lengths of lenses L5 and L6.

The complex amplitude of the incident light $E_0(x, y)$ can be presented as

$$E_0(x,y) = A_0(x,y) \exp\left[-i\varphi_0(x,y)\right] \tag{1}$$

where $A_0(x, y)$ and $\varphi_0(x, y)$ are the amplitude distribution and the phase distribution (i.e. wavefront) respectively, and (x, y) represents the spatial coordinates of amplitude and phase of the complex amplitude.

Assuming the focal length of L6 (i.e. f_6) larger than the focal length of L5 (i.e. f_5), the radial shearing ratio $s = f_6/f_5$ is larger than 1. Then $E_1(x, y)$ and $E_2(x, y)$, which are the complex amplitudes of two exiting beams from the cyclic RSI, can be presented by

$$\begin{cases} E_1(x,y) = A_0(x/s, y/s) \exp\left[-i\varphi_0(x/s, y/s)\right] \\ \qquad\qquad = A_1(x,y) \exp\left[-i\varphi_1(x,y)\right] \\ E_2(x,y) = A_0(xs, ys) \exp\left[-i\varphi_0(xs, ys)\right] \\ \qquad\qquad = A_2(x,y) \exp\left[-i\varphi_2(x,y)\right] \end{cases} \tag{2}$$

where $A_{1/2}(x, y)$ present the amplitude distributions of magnified and de-magnified beams respectively; $\varphi_1(x, y)$, $\Delta\varphi_2(x, y)$ presents the phase distributions of magnified and de-magnified beams respectively. Their definitions are shown as

$$\begin{cases} A_1(x,y) = A_0(x/s, y/s) \\ \varphi_1(x,y) = \varphi_0(x/s, y/s) \\ A_2(x,y) = A_0(xs, ys), \\ \varphi_2(x,y) = \varphi_0(xs, ys) \end{cases} \tag{3}$$

The aperture size of incident laser beam is limited as D by A1, and the corresponding beam size of exiting beams from cyclic RSI will be Ds and D/s respectively. The beam size is filtered again as same with their common area by A2 before they are introduced into FCCPS, as shown in **Figure 1**.

The FCPPS will generate the desired phase shifts of $\pi/2$ between each generated interferograms. Firstly, two beams, which generated from the cyclic RSI, are equally divided into two channels (i.e. channel a and channel b, shown in **Figure 1**) by the BS3. QW1 (angle distance of 0° from its fast axis to horizontal direction) and QW2 (angle distance of 45° from its fast axis to horizontal direction) are placed in channel *a*. QW1 generates a $\lambda/4$ phase delay between the Beam1 and the Beam2, and QW2 converts two beams into circularly polarized lights with levorotation and dextrorotation respectively. QW3 is placed into the channel b, and its fast axis keeps accordance to QW2. Another two circularly polarized beams with levorotation and dextrorotation are generated. The beams of channel a and b is split into four beam pairs by PBS2, i.e. a1, a2, b1 and b2 respectively, and it will introduces $\lambda/2$ phase delay into between channel 1 and 2. Four beam pairs are spatial arranged properly by mirrors M5 and M6 and beam splitter BS4, and they are projected onto the photosensitive plane of a CCD camera by the relay optics system composed of lenses L7 and L8. Finally, four interferograms are detected and recorded by the CCD camera.

The basic principle of the spatial phase-shifting can be explained by Jones formulas. The Jones matrices for a horizontal linear polarizer P_0, a vertical linear polarizer P_{90}, and a quarter-wave plate with horizontal fast axis Q_0, and 45° fast axis Q_{45}, are respectively

$$P_0 = \begin{bmatrix} 1 & 0 \\ 0 & 0 \end{bmatrix}, P_{90} = \begin{bmatrix} 0 & 0 \\ 0 & 1 \end{bmatrix}, Q_0 = \begin{bmatrix} 1 & 0 \\ 0 & -i \end{bmatrix}, Q_{45} = \frac{1}{\sqrt{2}} \begin{bmatrix} 1 & i \\ i & 1 \end{bmatrix} \tag{4}$$

The Jones matrix in the polarization state for each of the four channels can be represented by

$$\begin{aligned} a1 = P_0 Q_{45} Q_0 = \frac{1}{\sqrt{2}} \begin{bmatrix} 1 & 1 \\ 0 & 0 \end{bmatrix}, a2 = P_{90} Q_{45} Q_0 = \frac{1}{\sqrt{2}} \begin{bmatrix} 0 & 0 \\ i & -i \end{bmatrix}, \\ b1 = P_0 Q_{45} = \frac{1}{\sqrt{2}} \begin{bmatrix} 1 & i \\ 0 & 0 \end{bmatrix}, b2 = P_{90} Q_{45} = \frac{1}{\sqrt{2}} \begin{bmatrix} 0 & 0 \\ i & 1 \end{bmatrix} \end{aligned} \tag{5}$$

From Eq. (5), 0, 90, 180 and 270° phase shifting have been generated between each beam pair, which are the a1, b1, a2 and b2 respectively.

Wavefront difference $\Delta\varphi(x, y)$, which presents the difference between wavefront of Beam1 and Beam2 in their interference zone can be written as

$$\Delta\varphi(x,y) = \varphi_2(x,y) - \varphi_1(x,y) = \varphi_0(xs, ys) - \varphi_0(x/s, y/s) \tag{6}$$

The interference intensity of four interferograms, which are expressed by I_{a1}, I_{b1}, I_{a2} and I_{b2} respectively, can be represented by

$$\begin{aligned} I_{a1} &= A_1^2/2 + A_2^2/2 + A_1A_2\cos(\Delta\varphi) \\ I_{b1} &= A_1^2/2 + A_2^2/2 + A_1A_2\cos(\Delta\varphi + \pi/2) \\ I_{a2} &= A_1^2/2 + A_2^2/2 + A_1A_2\cos(\Delta\varphi + \pi) \\ I_{b2} &= A_1^2/2 + A_2^2/2 + A_1A_2\cos(\Delta\varphi + 3\pi/2) \end{aligned} \tag{7}$$

where I_{ai}, I_{bi}, A_i and $\Delta\varphi$ are the abbreviations of $I_{ai}(x, y)$, $I_{bi}(x, y)$, $A_i(x, y)$ and $\Delta\varphi(x, y)$, here i = 1,2.

Thus the wavefront difference $\Delta\varphi(x, y)$ can be calculated by classic four-frame phase shifting algorithm, which given by

$$\Delta\varphi(x,y) = \tan^{-1}\left[\frac{I_{b2}(x,y) - I_{b1}(x,y)}{I_{a1}(x,y) - I_{a2}(x,y)}\right] \tag{8}$$

Then, a suitable phase unwrapping algorithm is employed to acquire the absolute wavefront difference distribution.

2.2 Experimental results and analysis

As a comparison, a wavefront simulator (WS), which is composed mainly of an electrically addressed phase-only liquid crystal spatial light modulator (LC SLM), is adopt in experimental system to generate the candidate wavefront, and it will be measured by both a Hartmann-Shack wavefront sensor and the described RSI. For achieving this goal, the LC SLM should be adjusted in a phase-only modulation mode by keeping accordance of polarization state between polarizer and LC SLM. The generated wavefront by WS is measured separately, and compare with that obtained by the proposed RSI. The lenses L3 and L4 placed in HS WFS system relay the LC SLM to the HS WFS. The experiment is performed with a He-Ne laser, and its wavelength λ = 632.8 nm. The focal length of Lens1 and Lens2 are 250 mm and 300 mm respectively. Four interferograms are detected by one 8 bit, 576 × 768 pixels CCD camera. The full resolution of HS WFS is 32 × 32, and the aberrated wavefront are generated by a 512 × 512 pixel LC SLM.

Figure 2 shows the practical experimental setup of the proposed RSI. In experiments, we find the best angle for acquiring highest fringe visibility firstly. Then the LC SLM is controlled to generate a random aberration, which is a combination of 45-limit order Zernike polynomials. Four fringe patterns were detected simultaneously by the CCD camera, and they are shown in **Figure 3**.

The wrapped wavefront difference $\Delta\varphi(x, y)$ can be calculated according to Eq. (8). The unwrapped result is also shown in **Figure 4(b)**. A modal wavefront reconstruction method is employed to reconstruct the wavefront under test from the unwrapped wavefront differences. The RMS and the PV of the reconstructed wavefront are 0.4753λ and 3.3222λ, respectively, and here RMS and PV represent the root-mean-square value and peak-to-valley value of the wavefront under test. When we normalize the RMS value of each order Zernike polynomial to 1.0λ, the Zernike polynomial coefficients of the wavefront under test can be obtained by using the least square method, and it is shown in **Figure 5(b)**.

The same aberrated wavefront also detected by HS WFS separately. A spot array is detected is split into HS WFS, and the corresponding gradient distributions along horizontal and vertical directions can be calculated. Finally the wavefront under test was reconstructed again by using a similar modal reconstruction method. **Figure 6(a)**

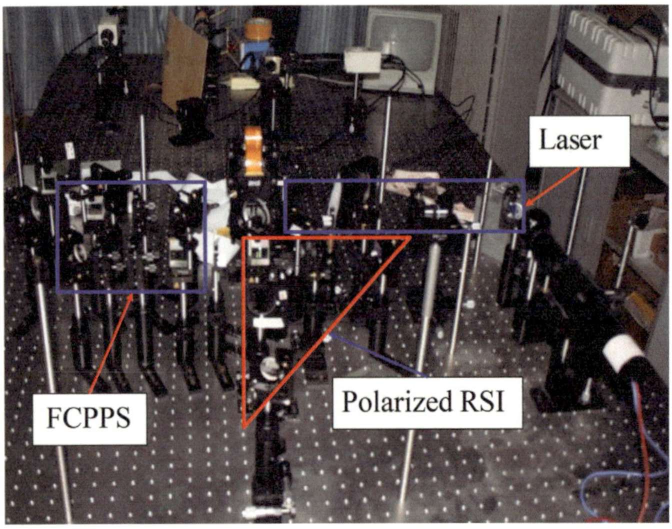

Figure 2.
Experimental setup of the proposed RSI.

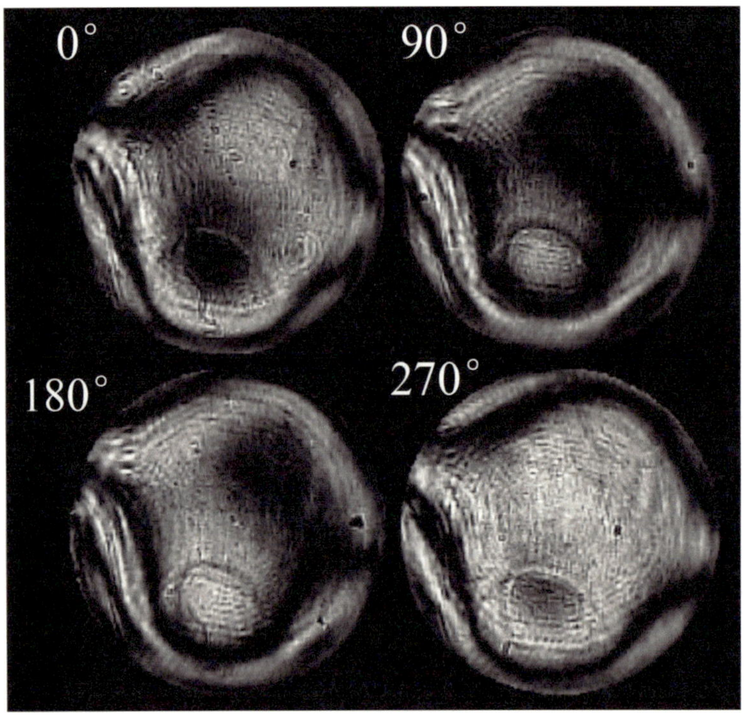

Figure 3.
Interferograms recorded by the experimental setup.

shows the reconstructed result, and the RMS and PV are 0.4798λ and 3.5953λ, respectively. The residual error, which is the difference between these two wavefront under test measured by the proposed RSI and the HS WFS, is also drawn in **Figure 6(b)**, and the corresponding RMS and PV are 0.0348λ and 0.3149λ, respectively.

From **Figure 6**, a small residual error is obtained, and the result measured by the proposed RSI keeping accordance with HS WFS. The difference is mainly due

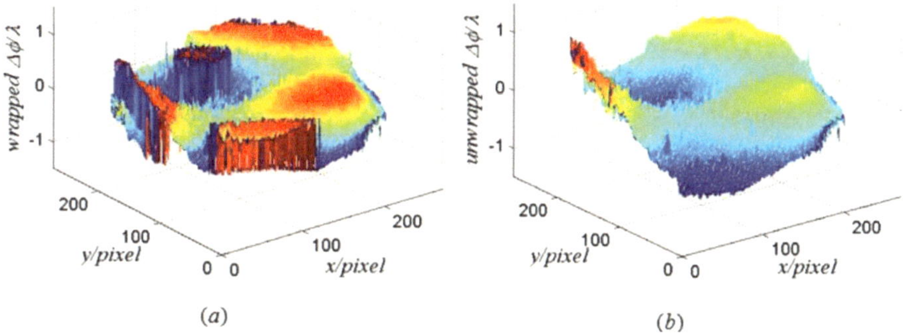

Figure 4.
The wavefront difference calculated from the four interferograms (shown in ***Figure 3****). (a) The wrapped wavefront difference and (b) the unwrapped wavefront difference.*

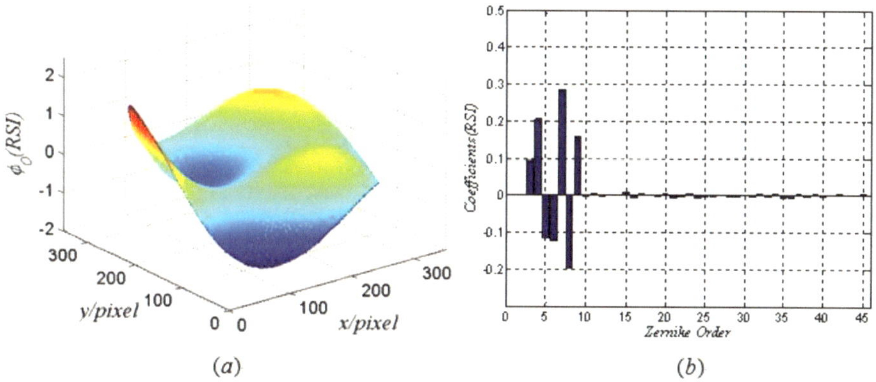

Figure 5.
The wavefront under test measured by the proposed RSI. (a) The 3D plot of the wavefront under test reconstructed from the wavefront difference shown in ***Figure 4(b)*** *and (b) the coefficients of the first 45 order Zernike polynomials.*

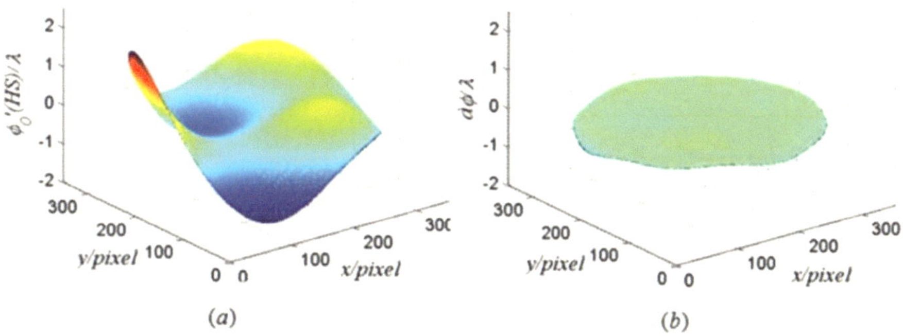

Figure 6.
The measured results of HS WFS and the residual error. (a) Experimental result measured by HS WFS (RMS and PV are 0.4798λ and 3.5953λ respectively) and (b) residual errors between wavefront measured by RSI and HS-WFS (RMS and PV are 0.0348λ and 0.3149λ respectively).

to the detecting noise of the CCD camera and the grating effect of the LC SLM. Better results would be acquired when a high performance CCD camera and a continuous surface wavefront simulator are used.

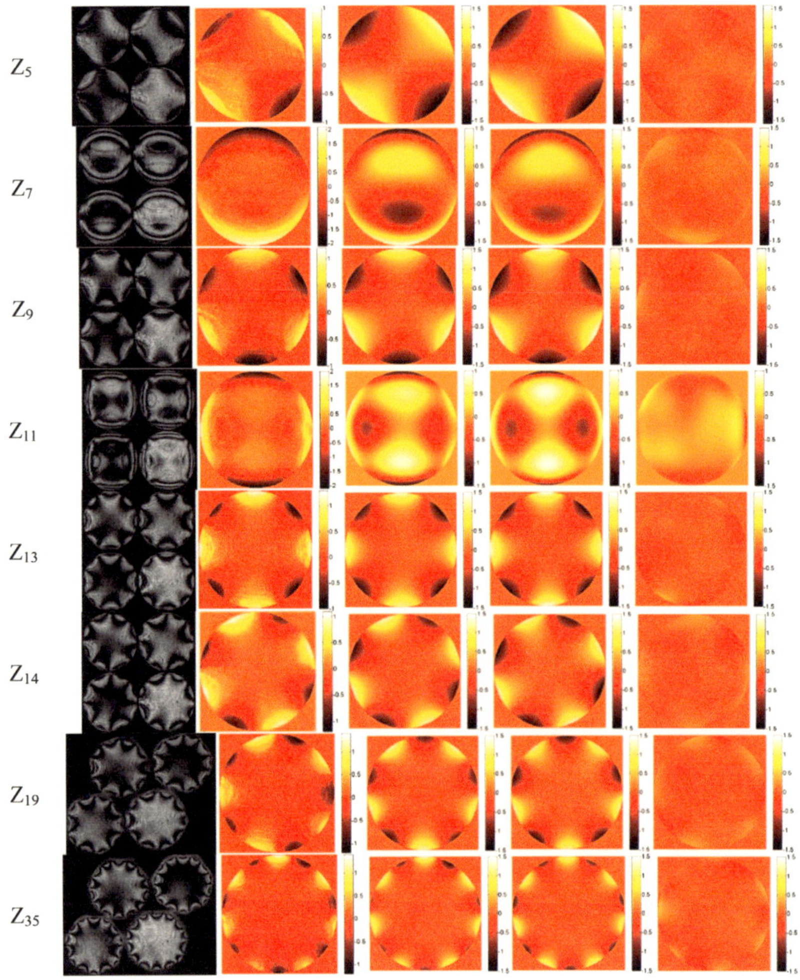

Figure 7.
Experimental results when the LC SLM generates some wavefront under test with single Zernike polynomial.

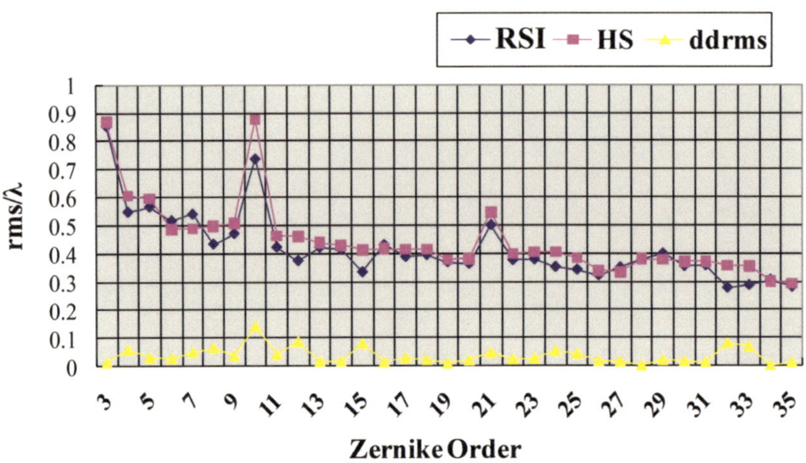

Figure 8.
Experimental results when the LC SLM generates some wavefront under test with single Zernike polynomial.

Figure 7 gives the other experimental results when the LC SLM generates some wavefront under test with single Zernike polynomial. **Figure 8** shows the statistic results of RMS and PV of 3–35 order Zernike polynomials measured by the proposed RSI and HS WFS.

3. Compact RSI based on micro-retarder array

In RSIs, complex configuration limits the measurement accuracy, system stability, and its application. Apparently, it is highly desirable to design compact one-shot vibration-insensitive RSIs. In this part, we will introduce a compact RSI based on pixelated micro-retarder-array (MRA).

3.1 Design and theory of the compact RSI

The basic concept and the schematic layout of the new compact RSI based on MRA are shown in **Figure 9**. The cyclic RSI is a Keplerian telescope system, which is same as the counterpart of **Figure 1**. The complex amplitude $E_0(x, y)$ can also be presented by Eq. (1). The half-wave retarder (HWR) is to adjust the polarization of the introduced light beam. Let θ as the angle between the polarization orientation of the introduced beam and the fast axis of the HWR. The polarization of the beam will be rotated by 2θ after passing through the HWR. The PBS splits the incident light into two beams with perpendicular polarization. The complex amplitude distributions of the new beam are generated when they pass through L1 and L2 and L2 and L1, respectively.

Figure 10 illustrates polarization conversion and complex amplitude after passing through different optical elements. $E_0(x, y)$ presents the optical field of the incident beam, and the variable α presents the angle of its SoP relative to x axis (i.e. horizontal direction). $E'_0(x, y)$ presents the complex amplitude of the exiting beam from HWR. The fast axis of HWR is presented by the dashed line. The complex amplitudes of reflection beam and transmission beam after passing through PBS can be presented as

$$E_p(x,y) = E'_0(x,y)\cos(\alpha - 2\theta), E_s(x,y) = E'_0(x,y)\sin(\alpha - 2\theta) \tag{9}$$

In fact the $E'_0(x, y)$ can be treated same as the $E_0(x, y)$ in scalar field because they have the same amplitude and phase distribution when they passing through HWR, which just change the SoP of the incident beam under different fast axis. Similar to the cyclic RSI in **Figure 1**, there are also two beams, i.e. magnified (i.e. $E_{exp}(x, y)$)

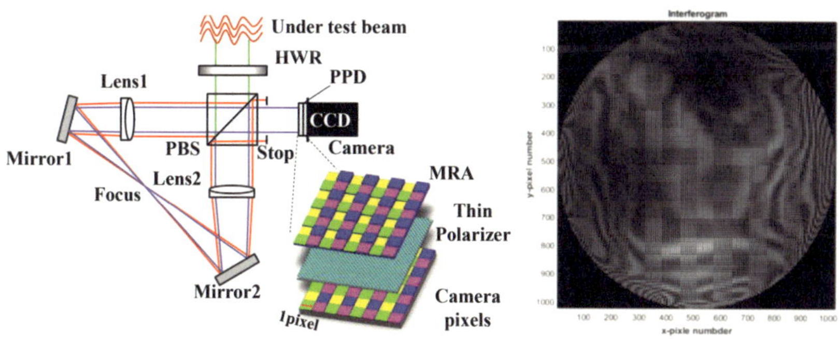

Figure 9.
Concept design of the proposed compact RSI. PPD, pixelated polarization device; HWR, half-wave retarder; PBS, polarized beam splitter; MRA, micro-retarder array.

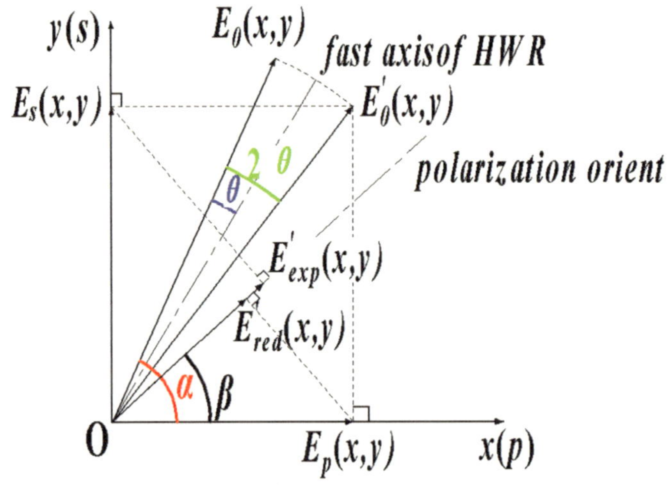

Figure 10.
Sketch map of SoPs conversion.

and de-magnified (i.e. $E_{red}(x, y)$) beams, are generated by the cyclic RSI, and they can be presented by

$$\begin{aligned} E_{\exp}(x,y) &= E_s(x/s, y/s)/s \\ E_{red}(x,y) &= sE_p(xs, ys) \end{aligned} \tag{10}$$

Finally the complex amplitude distribution of expanded and reduced beams can be presented as

$$\begin{aligned} E_{\exp}(x,y) &= \frac{1}{s}A_0'\left(\frac{x}{s},\frac{y}{s}\right)\exp\left[-j\varphi_0\left(\frac{x}{s},\frac{y}{s}\right)\right] \\ E_{red}(x,y) &= sA_0''(xs,ys)\exp\left[-j\varphi_0(xs,ys)\right] \end{aligned} \tag{11}$$

where $A'_0(x, y) = A_0(x, y)\sin(\alpha - 2\theta)$, and $A''_0(x, y) = A_0(x, y)\cos(\alpha - 2\theta)$.

Then the expanded beam and reduced beam with orthogonal polarization are introduced into the proposed phase-shifter, which consists of a micro-retarder array (MRA) and a piece of thin polarizer. Each macro pixel of MRA is composed of four neighboring pixels with different-thickness (as shown in **Figure 11(a)**). The MRA is made by a slice of birefringent crystal, and it can be quartz, calcite or the other kind of optical birefringent material. The fast axis of the selected birefringent crystal should be along or orthometric with x axis (i.e. horizontal direction) for keeping polarization status of incident light from changes when different phase-shifts are introduced. The schematic diagram of the MRA is drawn in **Figure 11**, and the basic principle of phase delay between two radial shearing beams is also introduced.

Pixel size of the MRA should be kept accordance with the pixel size of camera, and they should be integrated onto the photosensitive plane of the camera pixel by pixel. Here a negative single axis birefringent crystal is assumed, and let the fast axis of it along vertical direction. Thus the phase-delays of the four neighboring pixels is

$$\Delta\delta_i = (n_o - n_e)d_i \cdot 2\pi/\lambda, i = 1, 2, 3, 4 \tag{12}$$

where λ is the center wavelength of light, and i is the sequence number of the four neighboring pixels.

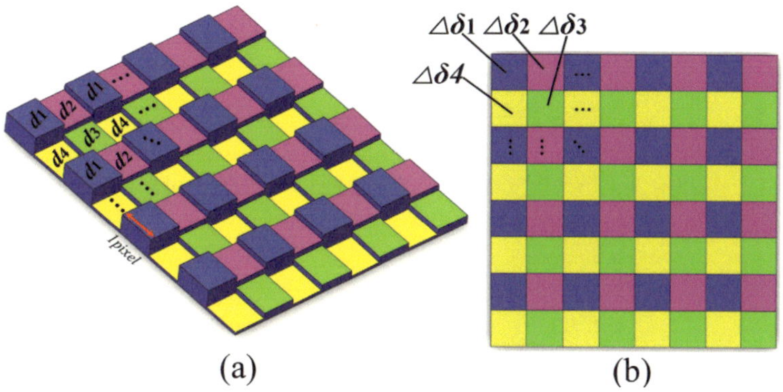

Figure 11.
Schematic diagram of the MRA. (a) MRA composed of units with different-thickness pixels made by a slice of birefringence crystal, and (b) the phase delay of each pixel. d_1–d_4*: thickness at four neighboring pixels,* $\Delta\delta 1$–$\Delta\delta 4$*: phase-shifting between* s- *and* p-*light for four neighboring pixels.*

When these two beams exiting from the cyclic RSI transmit through MRA, Eq. (11) at each pixel is changed as

$$\begin{aligned} E_{\exp,i}(x,y) &= A_0'(x/s,y/s)\exp\left[-j(\varphi_0(x/s,y/s)+\delta_i+\Delta\delta_i)\right]/s \\ E_{red,i}(x,y) &= sA_0''(xs,ys)\exp\left[-j(\varphi_0(xs,ys)+\delta_i)\right] \end{aligned} \tag{13}$$

where δ_i are the phase-shifts when the o-light transmits through MRA.

After passing through the thin polarizer, Eq. (13) can be rewritten by

$$\begin{aligned} E'_{\exp,i}(x,y) &= E_{\exp,i}(x,y)\sin\beta \\ E'_{red,i}(x,y) &= E_{con,i}(x,y)\cos\beta \end{aligned} \tag{14}$$

here β presents the filtering angle of the thin polarizer relative to horizontal direction, as shown in **Figure 10**.

After passing through the thin polarizer, the beams will be interference with each other on the pixels of the camera, and the intensity at four adjacent pixels can be represented as

$$\begin{aligned} I_i(x,y) &= \left|E'_{\exp,i}(x,y)\right|^2 + \left|E'_{red,i}(x,y)\right|^2 + 2\left|E'_{\exp,i}(x,y)\right|\left|E'_{red,i}(x,y)\right|\cos\left[\Delta\varphi(x,y)+\Delta\delta_i(x,y)\right] \\ &= I_a(x,y)+I_b(x,y)+2\sqrt{I_a(x,y)I_b(x,y)}\cos\left[\Delta\varphi(x,y)+\Delta\delta_i(x,y)\right] \end{aligned} \tag{15}$$

where $\Delta\varphi(x, y)$ represents the mean phase difference between the beams at (x, y) position. x and y presents the center position of each macro-pixel.

Combining the Eqs. (11), (13)–(15), a new equation can be obtained as

$$\begin{aligned} I_a(x,y) &= I_0\left(\frac{x}{s},\frac{y}{s}\right)\sin^2\beta\sin^2(\alpha-2\theta)/s^2, \\ I_b(x,y) &= s^2 I_0(xs,ys)\cos^2\beta\cos^2(\alpha-2\theta) \\ \Delta\varphi(x,y) &= \varphi_0(xs,ys)-\varphi_0\left(\frac{x}{s},\frac{y}{s}\right), \\ I_0(x,y) &= |E_0(x,y)|^2 = |A_0(x,y)|^2 \end{aligned} \tag{16}$$

Eq. (15) can be expressed in discrete form as

$$I_i(m,n) = I_a(m,n) + I_b(m,n) + 2\sqrt{I_a(m,n)I_b(m,n)}\cos\left[\Delta\varphi(m,n) + \Delta\delta_i(m,n)\right] \tag{17}$$

where (m, n) is a discrete form of (x, y), which presents the coordinate positions. Assuming the number of rows and columns of camera pixels are $2M_0$ and $2N_0$ respectively, the (m, n) should take a value from 1 to M_0 and 1 to N_0 respectively. Then Eq. (17) can be rewritten as

$$I_i(m,n) = K(m,n) + L(m,n)\cos\left[\Delta\delta_i(m,n)\right] - R(m,n)\sin\left[\Delta\delta_i(m,n)\right] \tag{18}$$

where

$$\begin{aligned} K(m,n) &= I_a(m,n) + I_b(m,n) \\ L(m,n) &= 2\sqrt{I_a(m,n)I_b(m,n)}\cos\left[\Delta\varphi(m,n)\right] \\ R(m,n) &= 2\sqrt{I_a(m,n)I_b(m,n)}\sin\left[\Delta\varphi(m,n)\right] \end{aligned} \tag{19}$$

Eq. (19) can be expressed by matrix form, and it is

$$\begin{bmatrix} I_1(m,n) \\ I_2(m,n) \\ I_3(m,n) \\ I_4(m,n) \end{bmatrix} = \begin{bmatrix} 1 & \cos\left[\Delta\delta_1(m,n)\right] & -\sin\left[\Delta\delta_1(m,n)\right] \\ 1 & \cos\left[\Delta\delta_2(m,n)\right] & -\sin\left[\Delta\delta_2(m,n)\right] \\ 1 & \cos\left[\Delta\delta_3(m,n)\right] & -\sin\left[\Delta\delta_3(m,n)\right] \\ 1 & \cos\left[\Delta\delta_4(m,n)\right] & -\sin\left[\Delta\delta_4(m,n)\right] \end{bmatrix} \begin{bmatrix} K(m,n) \\ L(m,n) \\ R(m,n) \end{bmatrix} \tag{20}$$

Thus

$$\begin{bmatrix} K(m,n) \\ L(m,n) \\ R(m,n) \end{bmatrix} = \begin{bmatrix} A_1(m,n) & B_1(m,n) & C_1(m,n) & D_1(m,n) \\ A_2(m,n) & B_2(m,n) & C_2(m,n) & D_2(m,n) \\ A_3(m,n) & B_3(m,n) & C_3(m,n) & D_3(m,n) \end{bmatrix} \begin{bmatrix} I_1(m,n) \\ I_2(m,n) \\ I_3(m,n) \\ I_4(m,n) \end{bmatrix} \tag{21}$$

where

$$\begin{bmatrix} A_1(m,n) & B_1(m,n) & C_1(m,n) & D_1(m,n) \\ A_2(m,n) & B_2(m,n) & C_2(m,n) & D_2(m,n) \\ A_3(m,n) & B_3(m,n) & C_3(m,n) & D_3(m,n) \end{bmatrix} = \begin{bmatrix} 1 & \cos\left[\Delta\delta_1(m,n)\right] & -\sin\left[\Delta\delta_1(m,n)\right] \\ 1 & \cos\left[\Delta\delta_2(m,n)\right] & -\sin\left[\Delta\delta_2(m,n)\right] \\ 1 & \cos\left[\Delta\delta_3(m,n)\right] & -\sin\left[\Delta\delta_3(m,n)\right] \\ 1 & \cos\left[\Delta\delta_4(m,n)\right] & -\sin\left[\Delta\delta_4(m,n)\right] \end{bmatrix}^{+} \tag{22}$$

The matrix $\boldsymbol{K}$, $\boldsymbol{L}$ and $\boldsymbol{R}$ can be solved for all units $M_0 \times N_0$ by

$$\begin{aligned} K &= A_1. \times I_1 + B_1. \times I_2 + C_1. \times I_3 + D_1. \times I_4 \\ L &= A_2. \times I_1 + B_2. \times I_2 + C_2. \times I_3 + D_2. \times I_4 \\ R &= A_3. \times I_1 + B_3. \times I_2 + C_3. \times I_3 + D_3. \times I_4 \end{aligned} \tag{23}$$

For constant phase-shifts $\triangle\delta_1$–$\triangle\delta_4$, A_1–D_3 are constant coefficients that can be calculated by Eq. (22) or calibrated by only one time. They can be used to extract random phase delay when the matrix of Eq. (22) is reversible.

Combining Eqs. (19) and (23), the phase difference between two shearing beams can be presented as

$$\Delta\varphi(m,n) = \tan^{-1}[R(m,n)/L(m,n)] \tag{24}$$

The wavefront under test $\varphi_0(x,y)$ can be reconstructed by Zernike mode [11] and iterative methods [30, 31].

The fringe visibility can be adjusted by changing the angle of HWR. Let ρ_0 as the mean power density of beam under test, the power density of the reduced beam ρ_p and the expanded beam ρ_s exiting the polarizer can be written by

$$\begin{aligned} \rho_p &= \rho_0 \cos^2(\alpha - 2\theta) \cdot s^2 \cdot \cos^2\beta \\ \rho_s &= \rho_0 \sin^2(\alpha - 2\theta)/s^2 \cdot \sin^2\beta \end{aligned} \tag{25}$$

For the best fringe visibility, ρ_p should be equal to ρ_s, and that means

$$\tan(\alpha - 2\theta)\tan\beta = s^2 \tag{26}$$

Generally, for a certain RSI system, the shear ratio s and the angle β are constants, so the fringe visibility can be adjusted by change of θ according to the polarization angle α of the beam under test.

3.2 Validation by numeric analysis

In this part, numeric analysis is made for validating the feasibility of the proposed compact RSI. The systematic parameters are listed as follows: aperture size Φ = 5 mm, random polarization angle of beam under test α = 139°, wavelength λ = 632.8 nm, focal lengths of Lens1 f_1 = 50 mm, focal lengths of Lens2 f_2 = 60 mm, β = 45°. And calcite retarder is chosen with vertical fast axis and (n_e, n_o) = (1.4852, 1.6557), thickness of four pixels per unit are [0.0591, 0.1536, 0.3072, 0.4785] μm. The pixel size of 1024 × 1024 camera is 9.8 μm × 9.8 μm. For the proposed RSI, the phase-shifts at four adjacent pixels per unit are [0.6283, 1.6336, 3.2673, 5.0894] rad, and θ is 41.89°for acquiring the best fringe visibility.

The beam under test is generated with random complex amplitude, which are shown in **Figure 12(a)** and **(b)**, respectively. The interferogram generated by the proposed RSI is calculated from Eqs. (12), (15) and (16), as shown in **Figure 12(a)**. The matrix ***K***, ***L*** and ***R*** are calculated by Eqs. (22) and (23). The wavefront difference between two shearing beams, as shown in **Figure 12(b)**, is extracted by Eq. (24).

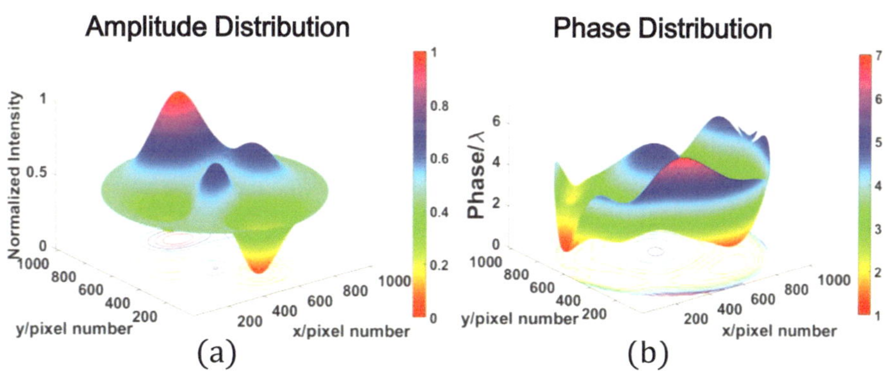

Figure 12.
Complex amplitude of the beam under test. (a) Random amplitude distribution generated by combination of 3-Gaussi functions, and (b) random phase distribution generated by combination of 65-limit Zernike polynomials (RMS = 0.975λ, PV = 5.736 λ).

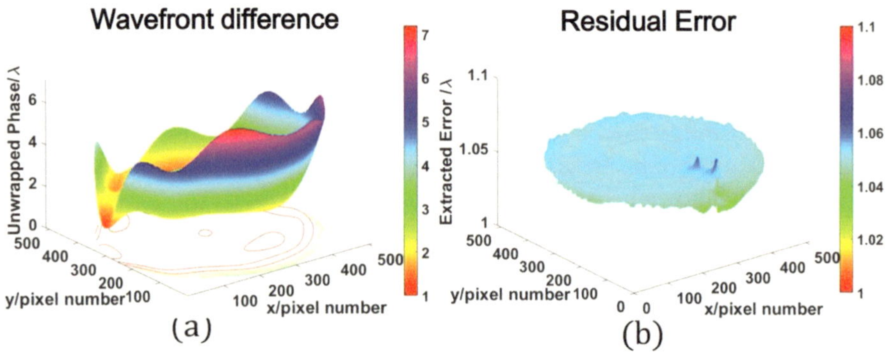

Figure 13.
The wavefront difference and its residual error relative to ideal value. (a) The unwrapped wavefront difference (RMS = 0.972λ), and (b) the error of the extracted wavefront difference (RMS = 0.002λ).

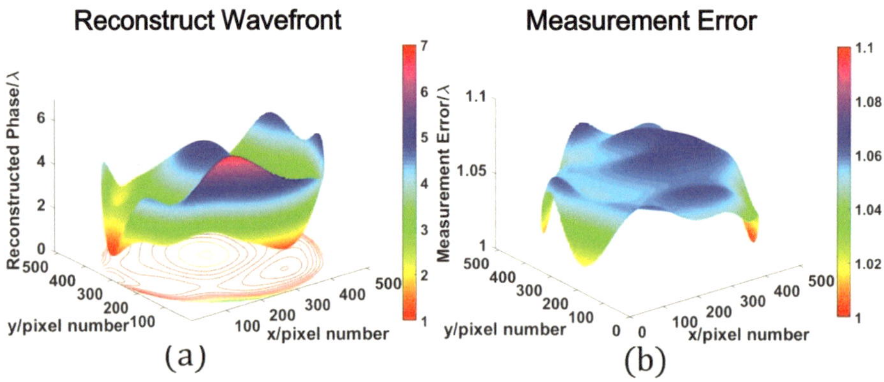

Figure 14.
The reconstruction result and the residual error. (a) The reconstruction result (RMS = 0.975λ), and (b) the residual error (RMS = 0.008λ).

Figure 13 shows the wavefront difference between two radial shearing beams and the deviation relative to the ideal wavefront difference. As seen from **Figure 13(b)**, the extracted error is less than 0.2%. The error is mainly caused by approximate treatment of intensity and phase values at four neighboring pixels. One should obtain a smaller deviation when a smoother intensity and phase distribution or large sampling number of camera is used.

Figure 14 shows the results of wavefront reconstruction. As seen from **Figure 14**, the RMS of the reconstructed wavefront is about 0.975λ, which is very close to the incident wavefront. The RMS of residual measurement error is about 0.008λ, which leads a relative deviation of 0.79%. In fact, if planar intensity amplitude and a smoother phase are employed in our numeric analysis, a better accuracy would be obtained based the method.

4. Modal method for reconstructing wavefront in RSI with or without later shear

In RSI no extra referencing planar wavefront but a magnified wavefront can be used as a reference. Thus the magnified and de-magnified beams should be aligned accurately avoiding wrong wavefront reconstruction. Unfortunately, two beams in practical application are often misalignment, and a lateral shear is always exists in a RSI. Here we present a modal method to solve this problem.

4.1 Modal reconstruction method based Zernike polynomials

As shown in **Figure 15**, the wavefront difference at interference area can be described as

$$\Delta\varphi(x,y) = \varphi_1(x,y) - \varphi_2(x,y) \tag{27}$$

The wavefront aberration $\varphi_1(x, y)$ is a de-magnified version of the wavefront under test $\varphi_0(x, y)$, and there no information is lost. So the wavefront aberration $\varphi_1(x, y)$ contains the same information with the wavefront under test $\varphi_0(x, y)$ if one normalize the diameter of the origin pupil and the de-magnified pupil. In other words, one can consider $\varphi_1(x, y)$ to be as the wavefront under test $\varphi_0(x, y)$. For simplicity, we define the interference area as $circle(d)$, and in this area two versions of wavefront under test can be expressed as

$$\begin{cases} \varphi_1(x,y) = \varphi_0(x,y) \\ \varphi_2(x,y) = \varphi_0(x/s^2, y/s^2) \end{cases}, (x,y) \in circle(d) \tag{28}$$

Eq. (28) is under an ideal alignment condition as shown in **Figure 15(a)**. However, in a practical RSI system lateral shear always exists and is inevitable. As shown in **Figure 15(b)**, the smaller circle denotes the de-magnified beam, and it also is the interference area, i.e. $circle(d)$. For convenience, the center of the contracted beam can be defined as the origin O of the Cartesian coordinate system, and the center position O′ of the magnified beam express the amount of lateral shear. Two variables, i.e. x_0 and y_0, are defined as the lateral shear at x direction and y direction respectively, and they can be calibrated by only one time measurement. Two versions of the wavefront under test can be rewritten as

$$\begin{cases} \varphi_1(x,y) = \varphi_0(x,y) \\ \varphi_2(x,y) = \varphi_0\left(x/s^2 - x_0, y/s^2 - y_0\right) \end{cases} \tag{29}$$

Suppose that the $\varphi_0(x, y)$ can be described by a N-limited Zernike polynomials, and thus $\varphi_0(x, y)$ can be expressed as

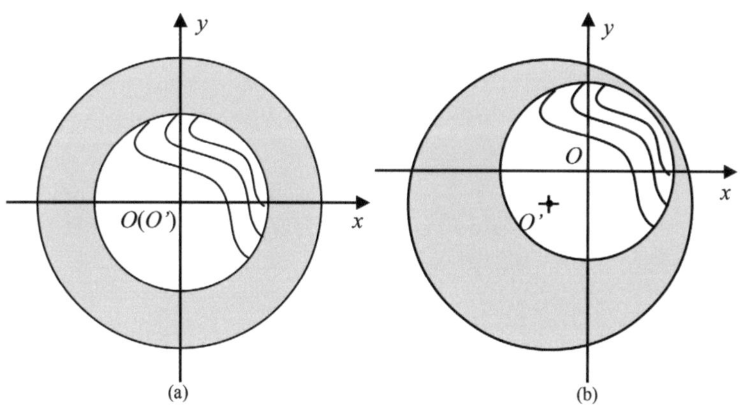

Figure 15.
Radial shearing interferogram without and with lateral shear. (a) Shearogram without lateral shear and (b) shearogram with lateral shear.

$$\varphi_0(x,y) = \sum_{k=0}^{N} a_k Z_k(x,y) \tag{30}$$

where $Z_k(x, y)$ is the kth order Zernike polynomial and a_k is the corresponding coefficient.

The magnified wavefront $\varphi_2(x, y)$ in *circle*(d) can be written by

$$\varphi_2(x,y) = \sum_{k=0}^{N} a_k P_k(x,y) \tag{31}$$

where $P_k(x, y)$ is a portion of Zernike polynomials $Z_k{}^D(x, y)$ in the expanded beam area (denoted as *circle*(D)), i.e. $P_k(x, y) = Z_k{}^D(x, y)$, $(x, y) \in circle(d)$.

According to the reference [32], provided an N-limited Zernike description of a wavefront on a large pupil, any circular portion inside it can be described by another Zernike ensemble, limited to the same N. So the function $P_k(x, y)$ can be expressed by

$$P_k(x,y) = \sum_{j=0}^{k} b_j^k Z_j(x,y) \tag{32}$$

Put the Eqs. (30)–(32) into Eq. (2), a new equation is generated, and it is

$$\Delta\varphi(x,y) = \sum_{k=0}^{N} a_k Z_k(x,y) - \sum_{k=0}^{N} a_k \left[\sum_{j=0}^{k} b_j^k Z_j(x,y) \right] \tag{33}$$

Eq. (33) describes the relationship between the wavefront difference and the Zernike polynomials. However, it is also indirect and not clear. Actually, Eq. (33) can be rewrite in its matrix form, and it is expressed by

$$\Delta\varphi = [a_1 \ \cdots \ a_k \ \cdots \ a_N] \left\{ \begin{bmatrix} 1 & & & & \\ & \ddots & & & \\ & & 1 & & \\ & & & \ddots & \\ & & & & 1 \end{bmatrix} - \begin{bmatrix} b_1^1 & & & & \\ \vdots & \ddots & & & \\ b_1^k & \cdots & b_k^k & & \\ \vdots & \vdots & \vdots & \ddots & \\ b_1^N & \cdots & b_k^N & \cdots & b_N^N \end{bmatrix} \right\} \begin{bmatrix} Z_1 \\ \vdots \\ Z_k \\ \vdots \\ Z_N \end{bmatrix} \tag{34}$$

where $\Delta\varphi$ and Z_k are the compact form of matrix formalism of $\Delta\varphi(x, y)$ and $Z_k(x, y)$, respectively.

Eq. (34) can be simplified as

$$\Delta\varphi = \mathbf{ABZ} \tag{35}$$

where $\mathbf{A} = \{a_1, \ldots, a_k, \ldots, a_N\}$ is a coefficient vector of the wavefront under test; $\mathbf{Z} = \{Z_1, \ldots, Z_k, \ldots, Z_N\}^T$ is a column vector which is composed by a set of Zernike polynomials; $\mathbf{B}$ is a coefficient matrix, and it is

$$\mathbf{B} = \begin{bmatrix} 1-b_1^1 & & & & \\ \vdots & \ddots & & & \\ -b_1^k & \cdots & 1-b_k^k & & \\ \vdots & \vdots & \vdots & \ddots & \\ -b_1^N & \cdots & -b_k^N & \cdots & 1-b_N^N \end{bmatrix} \tag{36}$$

On the other hand, the wavefront difference $\Delta\varphi(x, y)$ can be decomposed into a linear combination of the N-limited Zernike polynomials, and it is shown by

$$\Delta\varphi = \mathbf{CZ} \tag{37}$$

where $\mathbf{C} = \{c_1, \ldots, c_k, \ldots, c_N\}$ is the vector of coefficient.

There has only one unique decomposed result of Eq. (37) because the Zernike polynomials is orthogonal each other, and the solve can be presented as

$$\mathbf{C} = \mathbf{AB} \tag{38}$$

We can get

$$\mathbf{A} = \mathbf{CB}^{+} \tag{39}$$

where $\mathbf{B}^{+}$ is the generalized inverse of the matrix **B**.

The coefficient vector **C** can be calculated directly from Eq. (37), and the coefficient vector **A** can be determined by the matrix **B** finally. In fact, the matrix B can be determined beforehand by a numerical calculation or a calibration process for a certain radial ratio s and lateral shear x_0 and y_0. Then the wavefront difference $\Delta\varphi(x, y)$ can be expressed by a set of Zernike polynomials, and the corresponding coefficient vector **C** is calculated. Finally, the coefficient vector **A** can be determined by only one time matrix operation, and then the wavefront under test is reconstructed.

4.2 Numeric analysis

A random phase distribution is generated as wavefront under test for validating the proposed modal reconstruction method. A 45-limit Zernike polynomials are used, and the full resolution is 256 × 256. 1.2 of radial shear ratio s is assumed in a RSI, and η_x = 10% and η_y = −18% of relative lateral shear at x direction and at y direction respectively are used in our numeric analysis.

Here $\varphi_0(x, y)$ is presented by a random composition of 45-limited Zernike polynomials. Two variables, σ_{rms} and σ_{pv}, are used to express the root-mean-square (RMS) and peak-to-valley (PV) of the wavefront respectively. The 3D-plot of the simulated random wavefront $\varphi_0(x, y)$ is shown in **Figure 16(a)**. The RMS and PTV of the simulated wavefront under test are 2.000λ and 12.285λ respectively, where λ

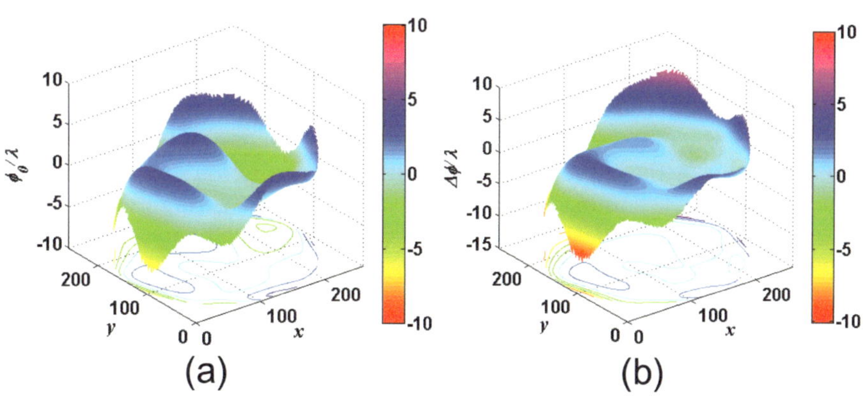

Figure 16.
Wavefront under test and the corresponding wavefront difference. (a) Random wavefront under test $\varphi_o(x, y)$ and (b) The wavefront difference $\Delta\varphi(x, y)$ calculated from $\varphi_o(x, y)$; the RMS and PTV of $\Delta\varphi(x, y)$ are 2.091λ and 18.057λ respectively.

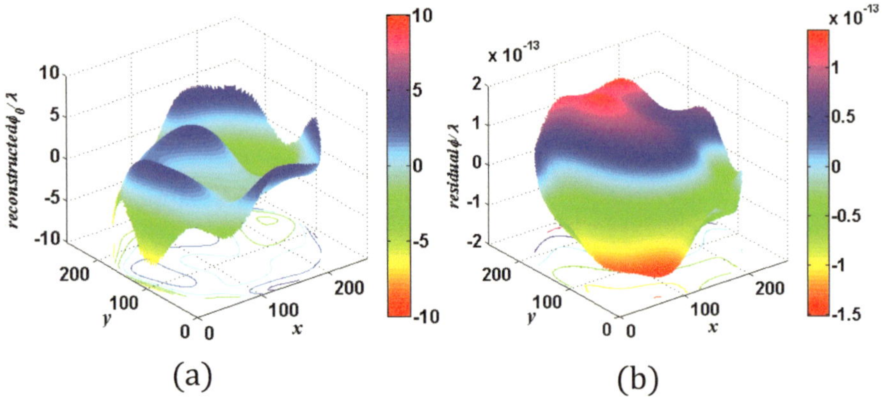

Figure 17.
*Wavefront reconstruction and the corresponding residual error. (a) The reconstructed wavefront under test by substituting the vector **A** into Eq. (30) and (b) the residual error between the origin wavefront and the reconstructed wavefront.*

is the wavelength of laser. According to the radial shear ratio s and the relative lateral shear η_x and η_y, the wavefront difference $\Delta\varphi(x, y)$ can be calculated easily, and it is shown in **Figure 16(b)**.

The coefficient matrix **B** can be calculated from Eqs. (32)–(36). According to the description of Eq. (39), the coefficient vector **A** is obtained. By applying Eq. (30), the wavefront of incident light can be reconstructed, and it is drawn in **Figure 17 (a)**. **Figure 17(b)** shows the residual error of wavefront reconstruction, and its corresponding RMS and PV are $7.003 \times 10^{-14}\lambda$ and $2.896 \times 10^{-13}\lambda$ respectively.

5. Conclusion

In this chapter, we have introduced briefly the history of RSI and the recent development. Two kinds of simultaneous RSI are also described in deep, including RSI based on simultaneous polarization phase-shifter and a new compact RSI based on micro-retarder array. Comparatively, the former one is suitable for wavefront measurements and diagnosing dynamic measurements, but its complex configuration limits the system stability, measurement accuracy and therefore its application. The new compact RSI based on micro-retarder array has several obvious advantages such as high-speed, high accuracy, vibration immunity, and compact size and so on, and it should be a kind of promising RSI in future. Complex extraction algorithm is another main barrier influencing the development of RSI. In this chapter a simple and useful modal reconstruction method is also given to extract the wavefront under test from the radial sheargrams with or without lateral shear. Comparing with the previous method, the modal wavefront reconstruction method reduces effectively the noise accumulation and has good error propagation property. It is based on Zernike polynomials and its matrix formalism, and it should lead to an easier implementation in some practical situations.

Acknowledgements

The authors are grateful to Prof. Wenhan Jiang for helpful discussions and suggestions. We would like to thank Prof. Changhui Rao, Prof. Linhai Huang, and Dr. Benxi Yao at Institute of Optics and Electronics (IOE), Chinese Academy of Science (CAS), for their cooperation and important help.

Funding

This work is supported by the National Natural Science Foundation of China under Grant Nos. 11643008 and 11727805.

Author details

Naiting Gu[1*] and Qun Luo[2]

1 Institute of Optics and Electronics, Chinese Academy of Sciences, Chengdu, China

2 National Key Laboratory of Science and Technology on Blind Signal Processing, Chengdu, China

*Address all correspondence to: gnt7328@163.com

References

[1] Hariharan P, Sen D. Radial shearing interferometer. Journal of Scientific Instruments. 1961;**38**:428-432. DOI: 10.1088/0950-7671/38/11/305

[2] Liu D, Yang Y, Shen Y. System optimization of radial shearing interferometer for aspheric testing. Proceedings of SPIE. 2007;**6834**: 68340U-1-8. DOI: 10.1117/12.760132

[3] Wang M, Zhang B, Shouping N. Radial shearing interferometer for aspheric surface testing. Proceedings of SPIE. 2002;**4927**:673-676. DOI: 10.1117/12.471690

[4] Kohno T, Matsumoto D, Yazawa T. Radial shearing interferometer for in-process measurement of diamond turning. Proceedings of SPIE. 1997;**3173**: 280-285. DOI: 10.1117/12.294519

[5] Kowalik W, Garncarz B, Kasprzak H. Corneal topography measurement by means of radial shearing interference: Part I—Theoretical consideration. Optik. 2002;**113**:39-45. DOI: 10.1078/0030-4026-00113

[6] Kowalik W, Garncarz B, Kasprzak H. Corneal topography measurement by means of radial shearing interference: Part II—experiment results. Optik. 2002;**113**:46-50. DOI: 10.1078/0030-4026-00112

[7] Kowalik W, Garncarz B, Kasprzak H. Corneal topography measurement by means of radial shearing interference: Part III—measurement errors. Optik. 2003;**114**:199-206. DOI: 10.1078/0030-4026-00247

[8] Gu N, Huang L, Yang Z. Changhui Rao: A single-shot common-path phase-stepping radial shearing interferometer for wavefront measurements. Optics Express. 2011;**19**:4703-4713. DOI: 10.1364/OE.19.004703

[9] Cheung D, Barnes T, Haskell T. Feedback interferometry with membrane mirror for adaptive optics. Optics Communication. 2003;**218**:33-41. DOI: 10.1016/S0030-4018(03)01188-X

[10] Shirai T, Barnes TH, Haskell TG. Adaptive wave-front correction by means of all-optical feedback interferometry. Optics Letters. 2000;**25**: 773-775. DOI: 10.1364/OL.25.000773

[11] Gu N, Huang L, Yang Z, Luo Q, Rao C. Modal wavefront reconstruction for radial shearing interferometer with lateral shear. Optics Letters. 2011; **36**:3693-3695. DOI: 10.1364/OL.36.003693

[12] Li D, Wang P, Li X, Yang H, Chen H. Algorithm for near-field reconstruction based on radial-shearing interferometry. Optics Letters. 2005;**30**: 492-494. DOI: 10.1364/OL.30.000492

[13] Liu D, Yang Y, Wang L, Zhuo Y. Real time diagnosis of transient pulse laser with high repetition by radial shearing interferometer. Applied Optics. 2007;**46**:8305-8314. DOI: 10.1364/AO.46.008305

[14] Hernandez-Gomez C, Collier JL, Hawkes SJ, Danson CN, Edwards CB, Pepler DA, et al. Wave-front control of a large-aperture laser system by use of a static phase corrector. Applied Optics. 2000;**39**:1954-1961. DOI: 10.1364/AO.39.001954

[15] Murty M, Compact Radial A. Shearing interferometer based on the law of refraction. Applied Optics. 1964; **3**:853-858. DOI: 10.1364/AO.3.000853

[16] Murty M, Shukla RP. Radial shearing interferometers using a laser source. Applied Optics. 1973;**12**: 2765-2767. DOI: 10.1364/AO.12.002765

[17] Ru Q-S, Ohyama N, Honda T, Tsujiuchi J. Constant radial shearing interferometry with circular gratings. Applied Optics. 1989;**28**:3350-3353. DOI: 10.1364/AO.28.003350

[18] Silva DE. Talbot interferometer for radial and lateral derivatives. Applied Optics. 1972;**11**:2613-2624. DOI: 10.1364/AO.11.002613

[19] Smartt RN. Zone plate interferometer. Applied Optics. 1974;**13**: 1093-1099. DOI: 10.1364/AO.13.001093

[20] Mohanty RK, Joenathan CJ, Sirohi RS. High sensitivity tilt measurement by speckle shear interferometry. Applied Optics. 1986;**25**:1661-1664. DOI: 10.1364/AO.25.001661

[21] Joenathan C, Torroba R. Simple electronic speckle-shearing-pattern interferometer. Optics Letters. 1990;**15**: 1159-1161. DOI: 10.1364/OL.15.001159

[22] Paterson C, Notaras J. Demonstration of closed-loop adaptive optics with a point-diffraction interferometer in strong scintillation with optical vortices. Optics Express. 2007;**15**:13745-13756. DOI: 10.1364/OE.15.013745

[23] Bai F, Rao C. Experimental validation of closed-loop adaptive optics based on a self-referencing interferometer wavefront sensor and a liquid-crystal spatial light modulator. Optics Communication. 2010;**283**: 2782-2786. DOI: 10.1016/j.optcom.2010.03.032

[24] Bryngdahl O. Reversed-radial-shearing interferometry. Journal of the Optical Society of America. 1970;**60**: 915-917. DOI: 10.1364/JOSA.60.000915

[25] Mihaylova E, Whelan M, Toal V. Simple phase-shifting lateral shearing interferometer. Optics Letters. 2004;**29**: 1264-1266. DOI: 10.1364/OL.29.001264

[26] Kothiyal MP, Delisle C. Shearing interferometer for phase shifting interferometry with polarization phase shifter. Applied Optics. 1985;**24**: 4439-4447. DOI: 10.1364/AO.24.004439

[27] Chung CY, Cho KC, Chang CC, Lin CH, Yen WC, Chen SJ. Adaptive-optics system with liquid-crystal phase-shift interferometer. Applied Optics. 2006; **45**:3409-3414. DOI: 10.1364/AO.45.003409

[28] Naik DN, Ezawa T, Miyamoto Y, Takeda M. 3-D coherence holography using a modified Sagnac radial shearing interferometer with geometric phase shift. Optics Express. 2009;**17**: 10633-10641. DOI: 10.1364/AO.17.010633

[29] Gu N, Yao B, Huang L, Rao C. A compact single-shot radial shearing interferometer with random phase-shift. Optics Letters. 2017;**42**:3622-3625. DOI: 10.1364/OL.42.003622

[30] Lago E, Fuente R. Amplitude and phase reconstruction by radial shearing interferometry. Applied Optics. 2007; **47**:372-376. DOI: 10.1364/AO.47.000372

[31] Li D, Wen F, Wang Q, Zhao Y, Li F, Bao B. Improved formula of wavefront reconstruction from a radial shearing interferograms. Optics Letters. 2008;**33**: 210-212. DOI: 10.1364/OL.33.000210

[32] Ragazzoni R, Marchetti E, Rigaut F. Modal tomography for adaptive optics. Astronomy and Astrophysics. 1999;**342**: L53-L56

Chapter 3

Differentiating the Superposition Principle from the Measurable Superposition Effects in Interferometry

ChandraSekhar Roychoudhuri

Abstract

The *physical interaction processes* behind the emergence of dark and bright fringes registered by the detectors at the output of optical interferometers is explained. This knowledge should be helpful to interferometrists to make better physical interpretations of their data. The *belief* in mysterious "interference of single indivisible photon" will disappear once we recognize that the spatial or temporal energy re-distributions are generated by the physical transformation experienced by the detecting molecules drawing energy from all the light beams. The molecules could be photodetectors at the interferometer output, or the beam combining dielectric boundary. The superposition principle (SP), represented by the *linear* mathematical sum of two or more wave amplitudes, does not represent an observable phenomenon. The superposition effect (SE), represented by the *non-linear* square modulus of the joint dipolar stimulation of the detectors by all the superposed waves, is observable. We present two laboratory experiments to clarify these points. Both classical beam combiners and quantum detectors are capable of generating superposition fringes of intensity variations. The logic of "quantumness" of light is narrowly relevant only when a quantum detector deciphers the fringes; it is not valid for classical beam combiners. We will also discuss "entanglement" based on these experiments.

Keywords: superposition principle, superposition effect, single photon interference, response of detecting dipoles, Mach-Zehnder interferometer

1. Introduction

The objective of this chapter is to use a simple classical Mach-Zehnder interferometer separately under *spatial-fringe-mode* (Poynting vectors noncollinear in the output) and *scanning-fringe-mode* (Poynting vectors perfectly collinear in the output). The purpose is to demonstrate that "single photon interference" is a noncausal interpretation of the light-matter interaction process, where light energy absorbed by a detector array represents bright fringes and nonabsorption represents dark fringes. The process is not due to some mystical arrival and nonarrival of "photons" that defies the diffractive light propagation physics. The light propagation physics cannot remain strictly valid everywhere except in human-constructed

IntechOpen

interferometers and Young's double-slit diffractometers. We are using mathematical formalism, which we have accepted for a couple of centuries. We support our arguments based upon *strict causality built into our mathematics* and corroborated by measured data. This chapter considers light-matter interaction process exclusively from the standpoint of semi-classical model, where light is treated as classical electromagnetic waves and light detecting atoms and molecules as quantum entities that have discrete quantum mechanical binding energy levels (or bands) with finite transition probabilities.

In Sections 2.1 and 2.2, we use simple Mach-Zehnder interferometers with the Poynting vectors of the two combined output beams under (i) noncollinear and (ii) strict collinear conditions. The experiments demonstrate that the two incident signals from the two opposite sides of the beam combiner *must be simultaneously present to generate the observed superposition effects*. The simple mathematical formalism tells us that wave amplitudes by themselves never re-arrange their spatial or temporal energy distribution. They continue to propagate unperturbed by each other's physical presence. We have called this wave property, noninteraction of waves (NIW). In Section 3, we show a simple derivation of the Einstein's photoelectric equation as a superposition effect congruent with the semi-classical model. Since the basic mathematical formalism of Quantum Mechanics (QM) is correct, we must learn to avoid being trapped in the century-old and *un-necessary* circular arguments behind diverse mystical interpretation of Quantum Mechanics (QM). In Section 4, we describe this briefly by underscoring that the so-called "Measurement Problem" of QM is a fundamental "Information Retrieval Problem" out of all experiments, which cannot be solved by promoting elegant and beautiful mathematical theorems. It will require consistent and iterative attempts on the part of generations of scientists to visualize the invisible interaction processes going on in nature. Experimental data driven "evidence-based science" is the best approach to unravel nature's causal interaction processes while guided by strictly causal mathematical rules. Unfortunately, most of the details of the micro interaction processes are still beyond our capability to visualize directly. The problem is further compounded by the fact that the beautiful mathematical logics are inventions by the "subjective" neural networks of the human species.

2. Differentiating the superposition principle from the measurable superposition effect

2.1 Formulating the basic superposition equation

The most neglected issue in current books and literature is that the co-propagating and cross-propagating wave amplitudes pass through each other completely unperturbed (uninfluenced) by each other's presence in the absence of interacting materials. In other words, the superposed wave fronts by themselves do not generate observable interferometric fringes whenever they are superposed. This is noninteraction of wave (NIW) amplitudes [1]. Alhazen observed this phenomenon almost 1000 years ago using a set of candles and a pinhole camera [2]. Huygens underscored this in his book [3] around 1667. This is why it is important to remember that even quantum electrodynamics acknowledges that photon-photon interaction cross section is immeasurably small [4]. In fact, Dirac mathematically found that "different photons do not interfere with each other" (NIW?). Unfortunately, he introduced the noncausal notion that "a photon interferes only with itself" [5]. This assertion is noncausal because interference fringes always appear as some *physical transformation in a detector* induced by more than one amplitude signals carrying different phase information. Further, the dark fringes are not due to nonarrival of "photons." It is because the joint stimulations by the out-of-phase E-vectors (only when equal amplitude!) fail to stimulate the detecting dipoles, and hence, the field energy cannot enter into the detecting dipoles' *quantum cups*. We always represent the superposition equation by two separate amplitude terms, each containing its own phase factor representing separate and independent oscillations of the E-vectors. A single stable elementary particle (here, a "photon") could not be *multivalued* in its critical dynamic parameters at any single moment. Further, in the *absence of any other interacting force*(s), it should not, by itself, appear or disappear in some specific physical location for some specific instrumental alignments made by humans to generate dark and bright fringes. In writing supposition equation, we always completely ignore the mathematics necessary to represent inherent diffraction physics of light waves. This is only partially correct for collimated beams traveling only short distances. Under this assumption, we neglect spatial evolution of complex amplitudes of the two waves. Photons cannot independently change their trajectories. Propagation of EM waves is rigorously given by Maxwell's wave equation and Huygens-Fresnel diffraction integral. Noninteraction of photons or generalized noninteraction of wave (NIW) [1] is equivalent to the statement that the linear mathematical superposition principle (*SP) does not represent an observable phenomenon*. It is a correct mathematical starting point to derive the *observable superposition effect (SE)* as the *nonlinear square modulus* of the expression for the linear superposition principle. In spite of this existing knowledge, somehow we get systematic training to accept that individual "indivisible light quanta" interfere by themselves. The phrase, "indivisible light quanta," represents energy "hν," a quantum cup of energy that is exchanged between quantized materials and classical waves. The mathematical expression *"hν" is devoid of phase information*. Note that our equation representing superposition phenomenon consists of *two or more superposed waves with different amplitudes and phases carried by separate waves.* Insertion of "bra" and "ket" symbols on to the complex amplitude terms fails to accommodate the physical properties behind *the spontaneous and perpetually diffractive EM waves* through any medium or instrument (evolving spatial amplitudes, phases, and Poynting vectors). This diffractive propagation property has been correctly captured by classical diffraction integrals, which form the very foundation of classical optics since its inception in 1817 by Fresnel [6].

In a typical two-beam interferometer, like Mach-Zehnder, the classical expression for the superposition of two replicated output plane waves, crossing through each other at a small angle, can be expressed as the traditional *Mathematical Supposition Principle*:

$$E(t,\tau) = E_1(t) + E_2(t+\tau) = a_1 e^{i2\pi\nu(t-\tau/2)} + a_2 e^{i2\pi\nu(t+\tau/2)} \quad (1)$$

In **Figure 1**, we have assumed that relative propagation delay along the two independent paths in the interferometer is zero by virtue of physical alignment when they intersect at the X = 0, the origin where the detector array is placed to register the fringes. The intensity variation is along the X-axis following the relative path delay, $\tau = 2(x \sin\theta/c)$.

Eq. (1) represents the well-known mathematical superposition principle (SP). SP is not directly observable. Generation of observable superposition effect (SE), or data, involves a series of complex physical interaction processes. The data are generated as some physical transformation triggered by energy exchange between some interactants in our chosen instrument. The interactants are photoresponsive material dipoles and EM waves. First, the material dipoles in the detector array

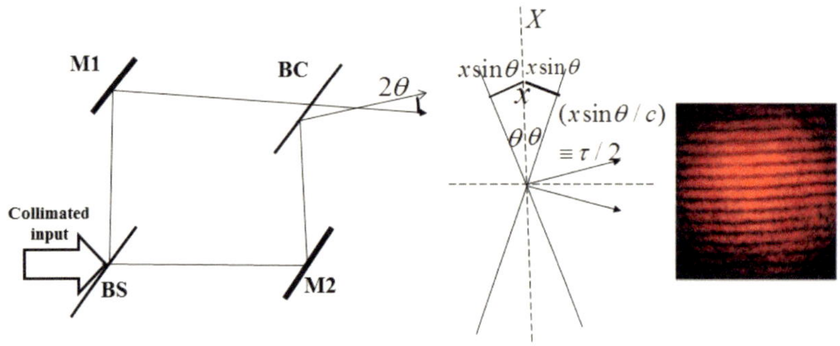

Figure 1.
A Mach-Zehnder interferometer in spatial-fringe-mode, illuminated with a collimated beam (left sketch). The two output beams intersect each other at the plane of a detector array (X-plane; middle sketch). The total relative path delay is $\tau = (2x\sin\theta)/c$. The linear dark-bright fringes oscillate along the X-axis (right photograph).

experience dipolar stimulations by both the superposed EM waves. The incident optical frequency ν has to be resonant with the allowed quantum transition. This *dipolar stimulation parameter* $\chi(\nu)$ is the *linear* (first order) polarizability. The traditional photo effects happen with the release of quantum mechanically bound photoelectrons to free space or internally to the conduction band. The older method used to be the dissociation of quantum mechanically bound Ag-Halide molecules in photographic plates. Therefore, the physical amplitude-amplitude stimulation effect, $\psi(t,\tau)$, should be explicitly recognized, as it is done by various semi-classical models [7–9]. Therefore, the real *physical superposition principle* should be expressed as the joint dipolar stimulations:

$$\begin{aligned}\psi(t,\tau) &\equiv \chi(\nu)E_{total}(t,\tau) = \chi(\nu)E_1(t) + \chi(\nu)E_2(t+\tau)\\ &= \{\chi(\nu)a_1\}e^{i2\pi\nu(t-\tau/2)} + \{\chi(\nu)a_2\}e^{i2\pi\nu(t+\tau/2)}\\ &= \chi(\nu)\left[a_1e^{i2\pi\nu(t-\tau/2)} + a_2e^{i2\pi\nu(t+\tau/2)}\right]; \text{frequency band very narrow.}\end{aligned} \tag{2}$$

(We are assuming here that the incident light beam is linearly polarized. Further, we are neglecting interactions of light with any anisotropic medium. This is to avoid unnecessary complexity in mathematics, which can divert our attention from the simple core issue that superposition effects become manifest only when some material medium undergoes observable physical change under the simultaneous joint stimulations by multiple waves.)

The interference fringe data $D(t,\tau)$ are generated *after the energy transfer process is executed* by the stimulated photosensitive dipoles, which is the square modulus of Eq. (2):

$$\begin{aligned}D(t,\tau) &\equiv |\psi(t,\tau)|^2 = \chi^2(\nu)\left[a_1^2 + a_2^2 + 2a_1a_2\cos 2\pi\nu\tau\right]\\ &= \chi^2(\nu)\left(a_1^2 + a_2^2\right)[1 + \gamma(\tau)\cos 2\pi\nu\tau]; \gamma(\tau) \equiv \left\{2a_1a_2/\left(a_1^2 + a_2^2\right)\right\}\\ &= 2\chi^2(\nu)a^2[1 + \cos 2\pi\nu\tau]; \text{only when } a_1 = a_2.\end{aligned} \tag{3}$$

This nonlinear square modulus interaction process cannot be executed by superposed linear fields in the absence of interacting materials. Accordingly, Eq. (1) does not represent any physical interaction process. *Therefore, we should restrain ourselves in assigning interpretations about the physical nature of light using this unobservable amplitude, Eq. (1), while, holding in mind, the observable data represented by the energy Eq. (3).* The proper linear physical superposition principle, representing joint amplitude stimulations of a detector, should always be represented by Eq. (2) to avoid making noncausal interpretations of observable superposition phenomena of nature. The best rational and physical interpretation that we can assign to Eq. (1) is that electromagnetic *waves can co-exist* within the same physical volume, while they cross-propagate, or co-propagate, independent of each other, as long as the medium is linear and noninteracting. This is the generalized noninteraction of waves (NIW), mentioned earlier valid for all propagating wave phenomena. For details, see Refs. [1, 10].

Let us appreciate that *the same single photon cannot represent two different physical beams.* Let us also note that in Eq. (3), the fringe visibility, $\gamma(\tau) \equiv \left\{2a_1a_2/\left(a_1^2 + a_2^2\right)\right\}$, is influenced by the two different amplitude parameters, and further, the fringe oscillation is influenced by two different phase factors, $\exp[i2\pi\nu(t-\tau/2)$ and $\exp[i2\pi\nu(t+\tau/2)$, which are manifest in the oscillation of the fringes through $\cos 2\pi\nu\tau$. We have mentioned that no *stable elementary particle* can carry more than one unique value for any of its physically essential (critical) dynamic parameters at any moment that define the existence of the entity. Thus, as per our causal equation, each one of the two signals we generate in an interferometer must be a physically *real and independent physical signal*, whether they are interpreted as classical EM wave packets or quantum mechanical "indivisible light quanta." Each wave plays separate important role in generating the final fringe pattern. Just one "photon," from one or the other wave, cannot dictate the emergence of the fringes as it does not have the other necessary parameter to interact with while stimulating the detecting dipoles. The concept of interaction-free superposition effect does not advance our intention of deeper enquiry of phenomena in nature. Natural phenomena happen only through diverse interactions between different entities.

Fringes of perfect unit visibility are almost impossible to achieve in practice. The fringe visibility is $\gamma(\tau) \equiv \left\{2a_1a_2/\left(a_1^2 + a_2^2\right)\right\} = 1$, only when $a_1 = a_2$ accurate to a tiny fraction of the energy carried by a single photon, $h\nu = \left(6.63 \times 10^{-34} J.\sec\right) \times \left(5.83 \times 10^{+14}/\sec\right) = 3.86 \times 10^{-19} J$ (for green light). This is a staggeringly small energy. We do not have any energy meter that can even come close to *directly* measure such a miniscule energy *decisively*. Further, the final beam combiner must have exactly T/R = (0.5/0.5), accurate to the single photon energy. A vendor will never agree to sell such a beam splitter, as they do not have the necessary technology to measure energy with such an accuracy. In other words, the fringe visibility, $\gamma(\tau) \equiv \left\{2a_1a_2/\left(a_1^2 + a_2^2\right)\right\}$, must be computed by separately measuring the two intensities, a_1^2 and a_2^2, for the two real beams and then compute a_1 and a_2. Eq. (3) has been experimentally validated innumerable times during the past two centuries. This experimentally valid mathematical Eq. (3) dictates that the superposition of two real and separate beams generates the interference fringes, and they cannot be generated by a single photon out of only one of the two beams.

Emergence of dark fringes: To strengthen the previous points further, we explore the physical processes behind the emergence of alternate dark fringes between the bright fringes. Mathematically, perfect dark fringes imply $D(t,\tau) = 0$. Superposition effect is *local* since the bright and dark fringes continuously vary in space and the size of the detecting dipoles is miniscule ($\sim$Angstrom^3) [11]. We all accept that the cosine intensity variation in the fringes is due to the phase factor $\cos 2\pi\nu\tau$, as we keep changing the relative path delay τ. $D(t,\tau)$ is a minimum whenever $2\nu\tau$ becomes an odd multiplier to π, making $\cos 2\pi\nu\tau = -1$. However, the only way $D(t,\tau)$ could be exactly zero is when $a_1 = a_2$, making the visibility factor $\gamma(\tau) = 1$. It is then obvious through the rigorous mathematical logic of Eq. (3) that dark fringes cannot be due to "nonarrival of indivisible photons" [12, 13] in these locations. Implication of "single photon interference" is that the detector's role is only to register the arrived energy without any participating in any way *in the superposition interaction process itself*. This noncausal assumption has led to the emergence of the *belief* behind *nonlocality*, *noncausality*, etc. This faithful belief in *nonarrival* of indivisible photons or of particles at dark fringe locations still dominates our physics community. It is then a fair question to ask whether we should trust more than a century old data validated Eq. (3) or the ***unnecessary*** mystical interpretation proposed after the advent of quantum mechanics in 1925. In the face of the reality that all theories are approximate and incomplete, we can keep on advancing science only by accepting any experimentally validated mathematical theories, which do not promote noncausal and incongruent interpretations. We should use such "evidence-based theories" *as intermediate steps* toward constructing better and better theories. In other words, QM is incomplete since it has been constructed to predict only the final measurable outcome *without any attention to understand the detailed interaction processes* that precede the QM transitions.

Most of the self-contradictory interpretations of the superposition principle will become unnecessary once we accept the hybrid model for photons. Atoms and molecules do emit discrete quantity of energy $h\nu$ during each allowed quantum transition. However, the $h\nu$ packet of energy immediately evolves into a perpetually propagating EM wave while spreading out diffractively, as it has been modeled classically (Ch. 10 in Refs. [1, 14]). During energy absorptions, the quantum dipoles function as "quantum cups" and they fill up that cup with the required amount of energy $h\nu$ out of the classical waves propagating through them. If the flux density within its stimulated quantum-cup volume is below $h\nu$, then there will be no photoelectric transition. This is a difficult experiment since one has to reduce the energy flow in reliable calibrated steps toward the single photon energy of $3.86 \times 10^{-19} J$. Current technology does not offer us any energy meter that can *directly* measure such a miniscule amount of energy. The release of a single bound photoelectron does validate the absorption of $h\nu$ amount of energy, since QM is essentially correct. However, it does not guarantee the existence and presence of *indivisible photons*. Each measured *click* out of our photon counting electronic system represents a burst of current pulse consisting of hundreds of millions of amplified electrons. As electrons are quantized particles and always bound quantum mechanically, their emission process steps will have to be discrete. Discreteness of the emitted electrons does not automatically validate the discreteness of the propagating EM wave energy. *Quantum formalism does not require that quantum transition can be affected only by another quantum entity having the right amount of energy to donate.* Striking stones create sparks. This is a complex energy exchange process, but nonetheless, it consists of a discrete series of quantum excitation and de-excitation processes triggered originally by the *mechanical energy of hands*. The validity of Boltzmann's rule of population density, dictated by classical *thermal collisions*, is rigorously obeyed by quantum systems that experience thermal collisions.

The author believes that nature abhors magic and mysticism. Physics has been advancing based on hard causality guided by rigorous mathematical logics but as a guiding tool only. Human invented mathematics cannot dictate nature how she ought to behave.

2.1.1 Recognizing noninteraction of waves (NIW) from the expression for the real physical superposition principle

We have already defined real *physical superposition principle* as Eq. (2), which represents the *linear amplitude* stimulation of the detecting dipole induced by both the superposed fields simultaneously. The energy transfer to the detector is given by the square modulus of Eq. (2), as presented in Eq. (3). Note that the polarizability $\chi(\nu)$ in Eq. (2) and $\chi^2(\nu)$ in Eq. (3) can be taken out of the amplitude and the intensity terms, respectively, as a common factor, provided χ does not vary with ν and t. This is allowed by the generic mathematical rule. This mathematical rule, while bears out in many situations, deprives us, appreciating the deeper interaction processes in nature. The t-independence of $\chi(\nu)$ is almost guaranteed whenever the material properties are stable in time. However, the frequency independence is never rigorously valid, since response of all materials depends upon the resonant frequency (quantum transition levels or bands in atoms and materials). We know that the detectivity of broadband solid-state detectors varies with frequency. However, when the spread of the wave frequency in the incident beam is extremely narrow, $\chi(\nu)$ can be treated as a constant for that frequency and treated as a constant common factor.

Let us now pay close attention to the third line of Eq. (2). With $\chi(\nu)$ taken out as a common factor, the expression within the parenthesis *may imply* that the two EM wave amplitude components are *summing themselves* to create the *interference* of waves. We can easily make this very serious conceptual mistake guided by the allowed abstract mathematical rules. Mathematical rules are human-invented logics. They do not always capture nature's logic. The linear superposition of amplitudes, as already underscored, does not generate any observable effect by themselves, in the absence of some interacting materials. This is why appreciating noninteraction of waves (NIW) is of such critical importance [1].

2.2 Pure "classical" interference (re-direction of energy) by a "passive" beam combiner

The interferometer arrangement we have utilized in the previous section (**Figure 1**) could be described as an interferometer in *spatial-fringe mode* [15, 16]. The Poynting vectors of the two propagating beams in **Figure 1** are *noncollinear*. In this section, we analyze an alternate arrangement where the Poynting vectors of the two pairs of output beams, generated by the output beam combiner BC, are perfectly collinear and coincident to each other. Such superposed parallel beams will continue to propagate with the same relative, but fixed, path delay generated within the interferometer (**Figure 2**). To keep the mathematical analysis simple, as in the last section, we are assuming that the relative physical path delay between the two beams is zero on arrival on the proper dielectric-coated surface of the BC. To

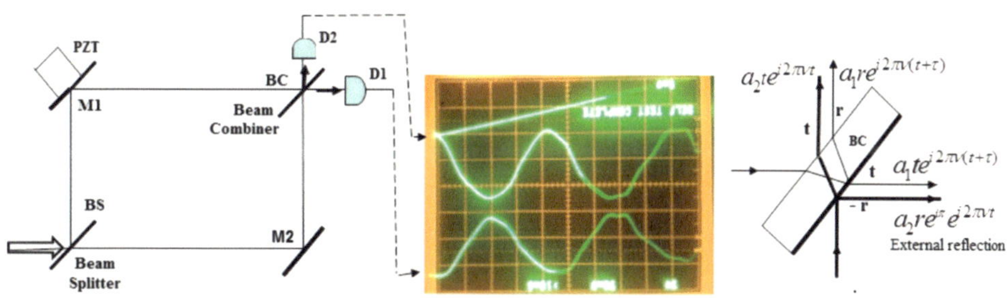

Figure 2.
A Mach-Zehnder interferometer in scanning-fringe-mode, illuminated with a collimated beam (left sketch). The Poynting vectors of the two pairs of output beams are perfectly collinear (right sketch). (For clarity of presenting mathematical expressions, the ray-traces are shown spatially separate. In real experiment, they are coincident.) The time-varying fringes can be observed for both the output ports D1 and D2 (middle photo), by scanning the mirror M1. The top-tilted linear trace in this photo represents the scanning voltage applied to the mirror M1. In this set up, the functional physical property of the 50% beam combiner BC oscillates between 100 and 0% transmittance (or reflectance). When D1 detector receives the maximum energy, D2 receives minimum energy [15, 16].

observe energy oscillating fringes under this condition, one has to scan one of the two mirrors with a piezo-electric drive to introduce the temporally oscillating path delay τ. Such interferometer arrangement, in *scanning-fringe-mode*, is quite common in today's lab.

As argued in the above section, the observable superposition effect can become manifest only through light-matter interaction. For a scanning-fringe interferometer, the interaction is between the dielectric boundary and the *two superposed wave amplitudes from the two opposite sides of this boundary material layer*. This is a pure classical boundary value problem for EM waves, which was actually solved by Fresnel even before the development of Maxwell's wave equations. Fresnel found that for the external reflection (towards the direction of lower refractive index), the wave amplitude suffers a π phase shift [1, 17]. The light-matter amplitude-amplitude stimulation is facilitated by the classical linear bulk polarizability $\chi(\nu)$ of the dielectric boundary materials. Here, this $\chi(\nu)$ is for bulk classical dipole polarizability. It is different from that $\chi(\nu)$ for dipoles undergoing QM transition, as in Eqs. (2) and (3), even though they are both *linear*. Then, the joint stimulation of the boundary layer induced by the right-going and the up-going beams, respectively, is:

$$\begin{aligned}\psi_{Rt.}(t,\tau) &\equiv \chi(\nu)E(t,\tau) = \chi(\nu)E_1(t+\tau) + \chi(\nu)E_2(t)\\ &= \{\chi(\nu)a_1 t\}e^{i2\pi\nu(t+\tau)} + \{\chi(\nu)a_2 r e^{i\pi}\}e^{i2\pi\nu t}\\ &= \chi(\nu)\left[a_1 t e^{i2\pi\nu(t+\tau)} - a_2 r e^{i2\pi\nu t}\right];\ \text{frequency band very narrow.}\end{aligned} \tag{4}$$

$$\psi_{Up}(t,\tau) = \chi(\nu)\left[a_1 r e^{i2\pi\nu(t+\tau)} + a_2 t e^{i2\pi\nu t}\right];\ \text{frequency band very narrow.} \tag{5}$$

Assuming that the dielectric boundary layer is approximately loss-less, we can assume that $\chi^2(\nu) = 1$. Then, the detectable energies in the right-going and up-going beams, respectively, are:

$$\begin{aligned}D_{Rt.}(\tau) &\equiv |\psi_{Rt.}(\tau)|^2 = \left|\left[a_1 t e^{i2\pi\nu(t+\tau)} + a_2 r e^{i\pi} e^{i2\pi\nu t}\right]\right|^2\\ &= \left[\left(a_1^2 t^2 + a_2^2 r^2\right) - 2a_1 a_2 tr\cos 2\pi\nu\tau\right]\\ &= a^2[1 - \cos 2\pi\nu\tau];\ \text{only if } r^2 = t^2 = 0.5 \text{ and } a_1^2 = a_2^2 = a^2\end{aligned} \tag{6}$$

$$\begin{aligned}D_{Up.}(\tau) &\equiv \left|\psi_{Up}(\tau)\right|^2 = \left|a_1 r e^{i2\pi\nu(t+\tau)} + a_2 t e^{i2\pi\nu t}\right|^2\\ &= \left[\left(a_1^2 r^2 + a_2^2 t^2\right) + 2a_1 a_2 tr\cos 2\pi\nu\tau\right]\\ &= a^2[1 + \cos 2\pi\nu\tau];\ \text{only if } r^2 = t^2 = 0.5 \text{ and } a_1^2 = a_2^2 = a^2\end{aligned} \tag{7}$$

From the very last lines of the energy, Eqs. (6) and (7), one can easily recognize that, under the ideal conditions of $r^2 = t^2 = 0.5$ and $a_1^2 = a_2^2 = a^2$, the sum total energy of the two beams, incident on the beam combiner from the opposite sides, remains constant, even though each output is oscillating between zero and $2a^2$:

$$D_{Rt..}(\tau) + D_{Up.}(\tau) = 2a^2,\ \text{with } a_1^2 = a_2^2 = a^2 \tag{8}$$

The implication is that a passive 50% beam splitter can dynamically keep changing its effective physical transmitivity, or reflectivity, between zero and one. The photo at the center of **Figure 2** is an experimental demonstration of this fact. A dual trace oscilloscope simultaneously displays the output of D1 and D2. While they are undergoing sinusoidal oscillation in displayed energy, the two traces are complementary to each other, preserving the rule of energy conservation. This simple classical *superposition effect* does not have any connection with quantum physics, since the beam splitter is purely a linear classical optical element. It is not undergoing any quantum transition or generating the fringes due to quantum absorption of energy. As mentioned before, the readers should note that currently we do not have technology to precisely set the conditions $r^2 = t^2 = 0.5$ and $a_1^2 = a_2^2 = a^2$ accurate to that of a *single photon energy*, $h\nu = 3.86 \times 10^{-19} J$. Accordingly, the claims that one can let a perfectly measured single photon incident on a beam combiner and then detect its appearance in either the port D1, or the port D2, must be taken with a "grain of salt." Recall that our counting ability of quantum mechanically released electrons does not make EM radiation quantized. Dirac's quantization of EM wave cannot *localize photons*; however, we have been generating highly localized, in time and space, laser energy as femto second pulses. The photons as linear superposition of finite or infinite set of Fourier modes are not a causal model of physics because of noninteraction of waves.

The conservation of energy remains true even when the conditions $r^2 = t^2 = 0.5$ and $a_1^2 = a_2^2 = a^2$ are not met. From the sum of the second lines of Eqs. (6) and (7), one can again derive that the total energy $(a_1^2 + a_2^2)$ is still conserved:

$$\begin{aligned} D_{Rt..}(\tau) + D_{Up.}(\tau) &= \left(a_1^2 t^2 + a_2^2 r^2\right) + \left(a_1^2 r^2 + a_2^2 t^2\right) \\ &= \left(t^2 + r^2\right)\left(a_1^2 + a_2^2\right) = \left(a_1^2 + a_2^2\right); \text{for loss-less beam splitter} \end{aligned} \tag{9}$$

From the standpoint of classical physics, the *π phase* shift for all *external reflection* is of critical significance behind the capability of a beam combiner to re-direct energy from one beam to the other. In fact, a homogeneous boundary layer always behaves like an "active anisotropic" layer because the boundary layer molecules are constrained to oscillate more easily as dipoles along the plane of the boundary, than orthogonal to the boundary, when an incident oscillating electric vector stimulates them. This is at the root of Brewster's Law, Malus's law, and partial polarization of "unpolarized" light in reflection. EM waves always stimulate the material dipoles to oscillate along the direction of its E-vector, while the wave propagates through any medium [17].

The π-phase shift in the external "reflection" (see the sketch for the BC in **Figure 2**, thick line from the bottom on the BC) means that the surface dipoles are oscillating in such a way as to allow only π-shifted light to be reflected compared to the incident beam. If a collinear "transmission" beam (**Figure 2**, thin line from the left on the BC) wants to pass through in the direction of the external reflection (to the right), it has to match this π-phase shift. However, if the phase of this left coming "transmission" beam happens to be zero or modulo-2π, its attempted dipole oscillation will be opposed by the "reflection" beam. The two incident amplitudes are now competing with each other in opposite directions. If the two amplitudes are equal, but opposite in phase, then all the energy of the two beams will be redirected along the upward direction, as *this direction provides an in-phase path for both the beams* and vice versa for the other direction.

The key point is that all beam combiners in all two-beam interferometers function this way when the Poynting vectors for the output beams are set for perfect collinearity and spatial coincidence. *The simultaneous physical presence of both the interfering beams from the opposite sides is must for the superposition effect to become manifest.* In other words, even if light consisted of "indivisible single photons," we must have the simultaneous presence of two photons from the opposite sides of the beam combiner. Hence, if "indivisible photons" were the reality, then such a classical interference effect would not have been observable with the incidence of truly "one photon at a time." The experiment we have presented used a 5-mW He-Ne beam, not "single photons." Such an experiment is being carried out in many senior level physics/optics laboratories.

The following challenge unequal-beam-energy experiment which is yet to be carried out. The superposition conditions can be dramatized even further. Even if one uses a beam combiner with $r^2 \neq t^2$ (the energy must be conservation rule is preserved, $r^2 + t^2 = 1$), one can then alter the incident energies a_1^2 and a_2^2 proportionately to completely suppress the energy propagation in *one of the two chosen direction* under the relative π-phase shift condition for that particular direction. *Would one be able to explain such an unequal-beam-energy experiment using the model of "indivisible single photon interference"?* Remember, the mathematical model consisting of the equations given above corroborates the measured data. Note that this generic mathematical model requires the *simultaneous* presence of the two signals from the two opposite sides on the beam combiner incident with the *right phases* and *right amplitudes.* Then only they can create the desirable transmission (or reflection) along either of the two output ports.

3. Re-deriving Einstein's photoelectric "energy" equation out of superposition effect

So far, we have only made passing negative comments about Einstein's 1905 of "indivisible light quanta" without presenting a better model to understand the interaction processes behind the photoelectron emission. In this section, we "derive" the photoelectric *energy* equation using the inevitable superposition of amplitudes of multiple wave groups. As before, we are following semi-classical model for light-matter interactions.

Classical thermal radiation consists of random wave groups emitted spontaneously with random phases. They keep propagating while diverging out due to waves' intrinsic diffraction property. The total dipolar amplitude stimulation of a bound electron can be expressed as:

$$\Psi = \Sigma_q \psi_q = \Sigma_q \chi(\nu_q) E(\nu_q) \tag{10}$$

The electron binding system must absorb the necessary $h\nu$ quantum-cup-filling energy, before the electron can be released to the conduction band, or become a free-space electron. This is a quadratic process:

$$|\Psi|^2 = \left|\Sigma_q \chi(\nu_q) E(\nu_q)\right|^2 \propto h\nu_q \tag{11}$$

We need to collect an *ensemble* of data to validate any characteristic behavior of a physical system. A single event (data point) is not sufficient to verify a theory.

$$\left\langle |\Psi|^2 \right\rangle = \left\langle \left|\Sigma_q \chi(\nu_q) E(\nu_q)\right|^2 \right\rangle \Leftrightarrow \langle h\nu_q \rangle = \left\langle \phi_{work\ fn.} + (1/2) m \mathrm{v}_{el.}^2 \right\rangle \tag{12}$$

Note that we have "recovered" Einstein's photoelectric *energy equation* out of dipole amplitude stimulations due to multitudes of waves. The left curve in Figure 3 was already published for Einstein to read [18]. His brilliance was to correlate the "minimum frequency" below which no photoelectron emitted to a unique "quantumness" in the phenomenon. He assigned this "quantumness" to "indivisible light quanta" in 1905 ("photon" word was coined in 1925 by G. N. Lewis). This was a brilliant idea at that time because even Bohr atom model with quantum mechanically bound electron was published 8 years later in 1913. However, *had Einstein correctly assigned the quantumness to bound electrons, he would have formulated*

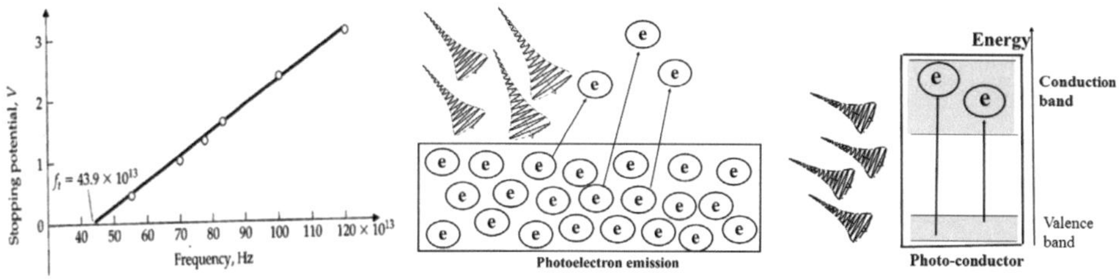

Figure 3.
Left diagram: Emission of photo electrons from a given material stops at a fixed specific frequency [18]. Middle diagram: Photoelectron emission from photo-cathode. Right diagram: Photoelectron transfer from valence to conduction band.

mathematics of quantum mechanics in his own style some 20 years earlier. Unfortunately, this neglect of the physical importance of amplitude stimulations as the critical first physical step before any quantum transition (energy exchange) can take place continued even after the QM was developed in 1925 and formalized during the next 10 years. We are stuck with the fictitious "indivisible light quanta" and "single photon interference." It would be a great cultural anthropological study to analyze and understand why the absolute majority *bend over backward* to justify Einstein's mistaken postulate, even though we now know that all electrons are bound quantum mechanically in materials. Recall the third line of Eq. (3). Waves only fill up the quantum cups with the necessary energy if the dipoles are resonant to the frequency ν of the incident waves (see left-diagram in **Figure 3**).

Let us rewrite Eq. (12) under the assumption that incident light is coming from a single frequency laser. In that case, $\chi(\nu_q)$ can be considered to be a constant and can be taken out of the summation sign:

$$|\psi_{res.}|^2 = \left|\Sigma_q \chi(\nu_q) E(\nu_q)\right|^2 = \chi^2 \left|\Sigma_q E(\nu_q)\right|^2 \tag{13}$$

The last step in Eq. (13) is meant to underscore again how some human invented mathematical logic, while corroborate measured data numbers can misguide us to believe that waves by themselves can carry out the complex operations of first (i) summing the wave amplitudes, and then (ii) taking square modulus of the sum.

4. Why we have been neglecting the obvious noninteraction of waves (NIW)

In both Sections 2 and 3, we have underscored the significance of noninteraction of waves (NIW). The neglect of NIW made us culturally oblivious to the fact that the *linear mathematical* superposition principle (SP) is not an observable phenomenon of nature. It *does not embody any interaction exchange process followed by nature*. In contrast, the superposition effect (SE) is an observable phenomenon because it models light-matter interaction process through the *nonlinear quadratic operation* $\psi^*\psi$. Note further that SE becomes manifest only in material detectors when their frequency resonant property matches with the frequency of the incident wave. NIW was known at least about 1000 years back [2]. It was repeatedly re-discovered but was neglected [10]. The puzzling issue is that this NIW is mathematically built into Huygens-Fresnel diffraction integral. Many modern undergraduate texts show constructions of Huygens' secondary wavelets, evolving through each other unperturbed, to model diffraction, reflection, and refraction, while propagating through linear and anisotropic media [17]. Yet, current books and literature never mention NIW. However, they present the linear superposition principle as the summation of wave amplitudes as if they physically sum themselves even while explaining that all propagation of light involves stimulating the dipoles of the material media. Note that the physical summation of wave amplitudes of different frequencies is behind the derivation of "group velocity," which is not an allowed physical process of nature. Because of the flexibility in interpreting math, our time-integrated data gathering practice, etc., many of the data validate the mathematical model. This is how we have *learned to accept* SP as an observable phenomenon by itself through years' of training. In the process, we have learned to suppress our inherent enquiring minds regarding the obvious contradictions while explaining many optical phenomena (see Ch. 1 in Ref. [1]). How have we acquired this culture of deferring to math and data without enquiring about the *physical interaction processes*? It took a series of grand mathematical successes in physics, during late 1800s and early 1900s, provided by only math and data, which is now known as the "evidence-based science." Evidence-based (experimental) science is the best possible approach to generate objective information about natural phenomena. In addition, mathematics is the best possible logical and objective tool, so far invented my humans, to model nature. Unfortunately, neither of them separately, or jointly, can provide us with the complete information about any interaction in nature that we study. We have learned to ignore the fact that, albeit math being the best objective logical language to model nature's behavior; it is just another human-invented language. Further, causal (or objective) mathematical equations require interpretations by human subjective minds. Math does not represent logical language that nature is obligated to follow. We have also learned to over-ride the fact that evidences (experimental data) still suffer from: (i) competing interpretations of data assigned by different subjective human neural networks, which have actually evolved for biological survival and not for the objective determination of the rules behind the biospheric and the cosmo-spheric evolutions. (ii) Further, no finite set of experimental data can provide us with the complete information about any natural entities. To be objective, we need to put careful and conscious efforts. Since the very beginning of biological evolution, our *pragmatic system engineering approach* to solve problems at hand with limited knowledge and trial-and-error has already made us the "top-dog" in the biosphere. We have installed the Global Internet System by figuring out the core system engineering functions for (i) generating, (ii) modulating, (iii) propagating, and (iv) detecting electrons and photons. However, we still do not understand what "photons" and "electrons" are made of. The lesson-to-be-learned is that a system engineering approach keeps us both grounded and pragmatically advancing. Whatever tools and technology works, they really obey the laws of nature even if we cannot articulate the rules.

Unfortunately, over the last century, we have deviated from consistently applying the system engineering approach to understand fundamental processes in nature [19]. This is evident from the broad cultural acceptance by the Knowledge Gatekeepers that the theories of Relativity and the mathematical formalism of Quantum Mechanics provide us with the final foundation of all future physics. Fortunately, the very long-term history of science does recognize that *all human constructed theories are necessarily incomplete since they have been formulated based upon insufficient knowledge of the universe.*

During the formative years of QM, people were marveled by the fact that Schrodinger's $\psi^*\psi$ always gives the right result even though they could not give proper physical explanation regarding what ψ physically represents. Born explained it as the abstract mathematical "probability amplitude," which must undergo an instantaneous "wave function collapse" to generate data in our instrument. This is certainly not a causal approach to understand and visualize the real physical *interaction processes* in nature, which gives rise to observable physical transformations. This conceptual gap between ψ and $\psi^*\psi$ gave birth to the concept of "measurement problem," which was then "solved" by using various elegant mathematical theorems. However, we have ignored the fact that if $\psi^*\psi$ gives us real world data, then ψ must also *contain* physical realities embedded in it. The author believes that ψ represents the real physical *amplitude stimulation* of the entity under study, as presented here in Sections 2 and 3 (see Ch. 3 in Ref. [1]). The square modulus operation represents energy exchanging *interaction process* between the interactants and between the energy donor and the energy recipient, guided by a force of interaction. These data generating interaction processes take a finite time. The interactants must first find the interaction *compatibility* with each other guided by the force they are experiencing and then undergo the full stimulation before jointly executing the square modulus operation to exchange the quantum cupful of energy. Let us elaborate the logical steps behind this physical process (see Ch. 12 in [1]).

i. *Data are physical transformations:* A properly designed instrument can register the anticipated data only when the measuring instrument receives the appropriate signal indicative of some physical transformation experienced by the chosen interactants.

ii. *Physical transformation require energy exchange:* Data generating physical transformation can happen only when the interactants exchange the necessary energy.

iii. *Some force of interaction guides the energy exchange:* Only an allowed force of interaction, compatible with the interactants, can guide energy exchange toward some physical transformation.

iv. *All "consumed" interactions are necessarily local:* Since all forces are of finite spatial range, the interactants must be physically present within this range of the guiding force. Thus, all interactions are "local" for the interactants as they must be within each other's vicinity to experience the mutual force, which "dies out" with distance. This implicates that interacting particles cannot remain "entangled" beyond the range of the effective force of interaction (in the sense of continued *mutually influencing* remote interaction). Nature is strictly causal, not mystical. There is no interaction free superposition effect.

The purpose of Physics is to figure out the invisible interaction processes that

i. *generate the data*: Our evolution and survival in the biosphere demand of us to think like system engineers and learn to manage the complex biospheric system. Unfortunately, we are not teaching our younger generation the extreme importance of understanding the depth of interaction processes going on in nature.

ii. *The information retrieval problem:* We can now appreciate that the validation of a theoretical model for any specific interaction by some experimentally measured data does not help us visualize the microscopic interaction processes that generated the data. This is an eternal "Information Retrieval Problem" for humans. This is not a "Measurement Problem" that can be solved by further refinement of the instrument or by inventing some elegant mathematical theorem. Nature is a creative system engineer. From this standpoint, we must figure out how to understand the details of the interaction processes. Today, we are facing the challenge of geoengineering the biosphere to slowdown the oncoming global warming. This is a formidable system-engineering task. Unfortunately, the physicists cannot contribute in this urgent process unless they are trained to enquire and understand the invisible interaction processes as system engineers.

iii. *How we can attempt to visualize the invisible interaction processes:* First, we need to change our culture to accept explicitly that "evidence based science" will always represent incomplete science. Only our consistent enquiry in understanding the interaction processes will allow us to start getting closer and closer to understand nature's system engineering marvels at both the micro and the macro levels. At present, we are happy with the Measurable Data Modeling Epistemology or MDM-E. MDM-E has successfully guided us to the highly advanced state we are in today. The author is proposing that we add Interaction Process Mapping Epistemology, or IPM-E, in conjunction with the prevailing MDM-E and accelerate our deeper understanding of nature as creative system engineers. Since the deeper realms and reality of nature are not directly visible to us, we will have to keep on applying IPM-E iteratively, perhaps, indefinitely. The progress will be slow, but we certainly will not fall in blind love with theories that are validated by data, but nonetheless do not represent nature's actual working rules. Slowly, we will become efficient and conscious sustainer of the biosphere, instead of continuing to be the plunderers. Declaring war against the evolutionary behavior of nature (outcomes of the laws of nature) will be futile until we master the extremely complex system engineering rules of the biosphere. This is why we must teach the students to keep on challenging the foundational postulates behind all working theories and assure consistent evolution in scientific theories. Sections 2 and 3 of this paper represent out of measurable data for superposition effects.

Let us recall Newton's philosophical view toward the end of his life: "If I have seen further than other men, it is by standing on the shoulders of giants." Since 1400 until today, we must have been accumulating new knowledge developed by many thousands of deep thinking scientists. Conceptually, we have built a pyramid of knowledge out of all these contributions. We should learn to climb to the apex of this pyramid to increase our knowledge horizon. We should not bow down our head out of "messiah complex" and reduce our knowledge horizon.

Einstein is known to have said: ".... After 50 years' of brooding over the question of what are light quanta; I still do not understand it!" This is the sign of a truly honest enquiring mind of a scientist. In spite of receiving the Nobel Prize for his proposed concept of "indivisible light quanta," he kept on questioning it due to causal inconsistencies. This public "questioning" attitude of Einstein has provided the author the strength to explore critically whether "indivisible light quanta" can really exist [20]!

5. Conclusions

Our key objective behind this chapter has been to re-introduce noninteraction of waves (NIW), which is a very important reality of nature in the field of interferometry. Since different "photon" amplitudes cannot interact even while co-propagating or cross propagating through each other, once spatially separate, *EM wave packets cannot keep on influencing each other remotely*; in this causal sense, they cannot remain "entangled." I leave it to the readers to make the necessary adjustments in their future thinking whether "single photon interference" truly represents the reality of nature [21, 22].

The readers should also note that the implication of the generic NIW-property of all waves is very significant in most of the branches of physics, some of which have been elaborated in my book [1]. I believe that incorporation of NIW in our analytical thought processes would help us develop deeper understanding of nature, besides guiding us to improve upon and/or invent many new instruments.

Author details

ChandraSekhar Roychoudhuri[1,2]*

1 Physics Department, University of Connecticut, Storrs, CT, USA

2 Femto Macro Continuum, Storrs, CT, USA

*Address all correspondence to: chandra.roychoudhuri@uconn.edu

References

[1] Roychoudhuri C. Causal Physics: Photon Model by Non-Interaction of Waves. Boca Raton: Taylor & Francis; 2014

[2] Ronchi V. Nature of Light. Cambridge: Harvard University Press; 1970. p. 53. See also page p. 18 in Ref. [1]

[3] Huygens C. Treatise on Light, drafted in 1678. Chicago: University of Chicago Press. Project Gutenberg edition; 2005. http://www.gutenberg.org/ebooks/14725

[4] Pike OJ, Mackenroth F, Hill EG, Rose SJ. A photon–photon collider in a vacuum hohlraum. Nature Photonics. 2014;**8**:434-436. http://www.nature.com/doifinder/10.1038/nphoton.2014.95

[5] Dirac PAM. The Principles of Quantum Mechanics. 4th ed. Oxford: Oxford University Press; 1974. p. 9

[6] Born M, Wolf E. Principles of Optics. Cambridge: Cambridge University Press; 1999

[7] Stroud CR, Jaynes ET. Long-term solutions in semiclassical radiation theory. Physical Review A. 1970;**1**(1): 106-121

[8] Barut AO, Van Huele JF. Quantum electrodynamics based on self-energy: Lamb shift and spontaneous emission without field quantization. Physical Review A. 1985;**32**(6):3187-3195

[9] Scully MO, Zubairy MS. Quantum Optics. Cambridge: Cambridge University Press; 1997

[10] Roychoudhuri C. Consequences of repeated discovery and benign neglect of non-interaction of waves (NIW). Proceedings of SPIE. 2017;**10452**: 1045215. Available from: https://www.spiedigitallibrary.org/conference-proceedings-of-spie [Accessed: 9/2/2017]

[11] Roychoudhuri C. The locality of the superposition principle is dictated by detection processes. Physics Essays, 2006;**19**(3):333

[12] Lande A. Quantum fact and fiction —IV. American Journal of Physics. 1975; **43**:701

[13] Mobley MJ. Momentum exchange theory of photon diffraction. Optical Engineering. 2018;**57**(1):015105. DOI: 10.1117/1.OE.57.1.015105.

[14] Roychoudhuri C. Hybrid photon model bridges classical and quantum optics. OSA Annual Conference, JW3A.32-1; 2017

[15] Roychoudhuri C, Barootkoob AM. Generalized quantitative approach to two-beam fringe visibility (coherence) with different polarizations and frequencies. Proceedings of the SPIE Conference. 2008;**7063**. paper #4

[16] Roychoudhuri C. Demonstration and implications when 50% beam combiners can behave as 0% or 100% reflector/transmitter inside some interferometers. Proceedings of SPIE. 2017;**10452**:04521C. This paper is free from the SPIE website

[17] Hecht E. See Section 4.6 in Optics. 5th ed. Reading: Addison Wesley; 2016

[18] Bernstein J, Fishbane PM, Gasiorowicz S. See Fig. 4.8 in Modern Physics. Upper Saddle River: Prentice Hall; 2000

[19] Roychoudhuri C. Urgency of evolution process congruent thinking in physics. Proceedings of SPIE. 2015; **9570**. paper# 7

[20] Roychoudhuri C. Most of the papers and presentations for workshop, etc., by

the author can be downloaded from: http://www.natureoflight.org/CP/

[21] Roychoudhuri C, Roy R, editors. The nature of light what is a photon? Optics & Photonics News (OSA). 2003; Special Issue. https://www.osa-opn.org/opn/media/Images/PDFs/3185_6042_30252.pdf?ext=.pdf

[22] Roychoudhuri C, Kracklauer AF, Kreath K. The Nature of Light: What Is a Photon? Boca Raton: Taylor & Francis; 2008

Chapter 4

Fiber-Optic Ring Resonator Interferometer

Ramón José Pérez Menéndez

Abstract

In this chapter, theoretical analysis and electro-optical characterization of a fiber optic passive ring resonator interferometer (FOPRRI) were realized. First, a theoretical description and analysis of all-pass and add-drop filter configurations were performed, hence obtaining their respective transfer function, taking into account, the physical properties of a standard single-mode optical fiber as resonator constitutive material. Second, computational numerical simulations of a typical FOPRRI were performed with different analysis methods: effective index method (EIM), coupled mode theory (CMT), finite element method (FEM) and finite difference time domain (FDTD). Third, a comparative analysis of the results obtained with these methods is realized, showing the most accurate and appropriate for filter electro-optical characterization as a consequence. Last, and as an ultimate step, taking as reference the previously obtained electro-optical characterization for both configurations, the main applications are derived.

Keywords: passive optical ring resonator interferometer, single-mode optical fiber, all-pass filter, add-drop filter, EIM (effective index method), CMT (coupled mode theory), FEM (finite element method), FDTD (finite difference time domain)

1. Introduction

Optical interferometry constitutes an important technique used in high number of measurement processes for multiple physical magnitudes and quantitative phenomenon [1]. Particularly, fiber-optic waveguides can act as very useful and efficient transmission medium for light guidance in many interferometric sensor devices. The issue of fiber-optic interferometric devices has received an extensive treatment in the literature [2, 3]. Also, the general theory of optical resonators was extensively treated by authors, for example, Yariv and Yeh [4]. However, quantitative analytic treatment of a fiber optic passive ring resonator interferometer (FOPRRI) can be rather found in singular applications, particularly, in gyroscopes. A clear proof of it is that one of the pioneering and singular descriptions of fiber-optic ring interferometer devices was in 1976 when Vali and Shorthill designed the first fiber-optic gyroscope model [5–7]. Later, many other authors studied in detail this first design and improved some of its electro-optical parameters [8, 9]. Considered from a broader point of view, the FOPRRI device has now evolved toward what we could call the integrated optical passive ring resonator interferometer (IOPRRI). In this case, the fiber-optic ring resonator is replaced by a ring-integrated optical waveguide, made of high refractive index material, such as lithium niobate, indium phosphide, silicon nitride, or silicon on insulator (SOI) platform. This chapter focuses on the analysis and testing of fiber-optic ring resonator. Therefore, the integrated ring resonator study, its investigations and challenges are beyond the scope of it [10].

An all-fiber passive ring resonator interferometer is formed by a directional coupler and a fiber-optic loop, as shown in **Figure 1**. Two main geometrical configurations are possible to reach such an optical coupling: (a) direct-coupled fiber ring resonator or (b) cross-coupled fiber ring resonator, as it can be seen in figure. Superimposition of successive beams generated by re-coupling the first coupled beam into the fiber ring results in multi-beam optical interference in a similar way to a Fabry-Pérot interferometer. Resonance will occur when the increase of the phase of the beam around the ring equals $2\pi n - (\pi/2)$, $\pi/2$ being the phase shift due to cross-coupling and n a positive integer (n = 1, 2, 3, ...), so that the light reentering the ring after a single circulation around it is in phase with the light entering the ring for the first time. Following L. F. Stokes et al. [12, 13], the transfer function of the system for the optimum amplitude resonance condition is given by:

$$\frac{I_T}{I_0} = (1-\gamma)\left\{1-\left[\frac{(1-\kappa)^2}{1+\kappa^2+2\kappa\,\sin(\beta L)}\right]\right\} \tag{1}$$

where γ is the insertion loss of the coupler, $\kappa = (1-\gamma)\ \exp(-2\alpha L)$ is the optical resonance coupling coefficient, α is the fiber attenuation factor, L is the total length of the fiber ring and β is the propagation constant of the re-circulating mode. The transfer function is very similar to a Fabry-Pérot interferometer operating in back-reflection mode. A finesse F of more of 1000 has been demonstrated for very low excess loss coupler and low-loss fibers [11, 12]. At resonance, significant optical power is recirculating in the ring and nonlinear effects as stimulated Brillouin scattering (SBS) can be observed for input powers of less than 100 μW [13]. As it will be seen later, fiber-optic ring resonators have been used in a number of applications, including fiber ring lasers [8], sensors [14], fiber laser gyroscopes [15, 16], optical spectrum analyzers [17, 18] and optical delay lines [19], among others. One of the first and main sensing applications of fiber-optic ring resonator is as a gyroscope and the major attraction of this kind of resonator gyro is that the cavity length can be taken much shorter than the convectional gyro in the ratio of 2/F. In order to detect the phase shift induced by rotation, a highly coherent laser light source is required. A dual frequency modulation scheme is used to control the laser frequency and tracking the resonance condition [8, 9]. The difference between

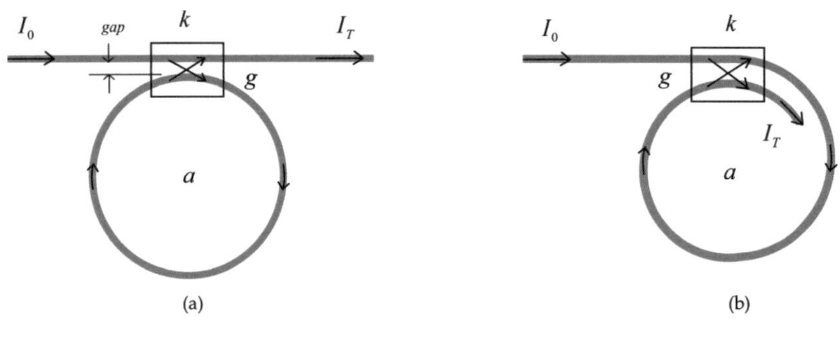

Figure 1.
Fiber-optic-passive-ring-resonator-interferometer (FOPRRI) all-pass configuration: (a) direct-coupled fiber ring-resonator and (b) cross-coupled fiber ring-resonator.

these two frequencies gives a measure of gyro output signal, which, in turn, is proportional to rotation rate to which the fiber-optic ring resonator system is subjected.

2. Analysis of the fiber-optic all-pass filter ring resonator configuration

The all-pass ring resonator configuration is created by taking one output of a generic 2 × 2 directional coupler and feedback into one input, as it can be seen in **Figure 2**. Such a device exhibits a periodic cavity resonance when light traversing the ring acquires a phase shift corresponding to an integer multiple of 2π radians. The resonator can mathematically be formulated from two components: a coupling strength (from coupling coefficient κ) and a feedback path (via transmission coefficient t), in contrast to the infinite sum derivations made for Fabry-Perot and Gires-Tournois interferometers. Therefore, following the text of Rabus [20], we will derive the basic spectral properties by assuming the steady-state operation and the matching fields. Although both methods are equally valid, the field-matching method has the advantage of simplicity. The basic relations among the incident E_{i1}, transmitted E_{t1}, and circulating E_{i2}, E_{it2} fields of a single resonator are derived by combining the relations for the coupler with that of the feedback path. In the frequency domain, the fields exiting the coupling region are related to the input fields via the following equations:

$$E_{t1} = t\,E_{i1} + i\,\kappa\,E_{i2} \tag{2}$$

$$E_{t2} = t\,E_{i2} - i\,\kappa\,E_{i1} \tag{3}$$

where factor i is the imaginary unit ($i = \sqrt{-1}$) that takes into account, the phase shift caused by the cross-coupling of the waveguides. For simplicity, we assume the total symmetry of cross-coupling and bar-coupling coefficients for both waveguides, and therefore, the two approximate equations are ($\kappa_1 \approx \kappa_2 = \kappa$ and $t_1 \approx t_2 = t$). The two Eqs. (2) and (3) can be written in a compact way via the following unitary matrix:

$$\begin{pmatrix} E_{t1} \\ E_{t2} \end{pmatrix} = \begin{pmatrix} t & i\kappa \\ -i\kappa & t \end{pmatrix} \begin{pmatrix} E_{i1} \\ E_{i2} \end{pmatrix} \tag{4}$$

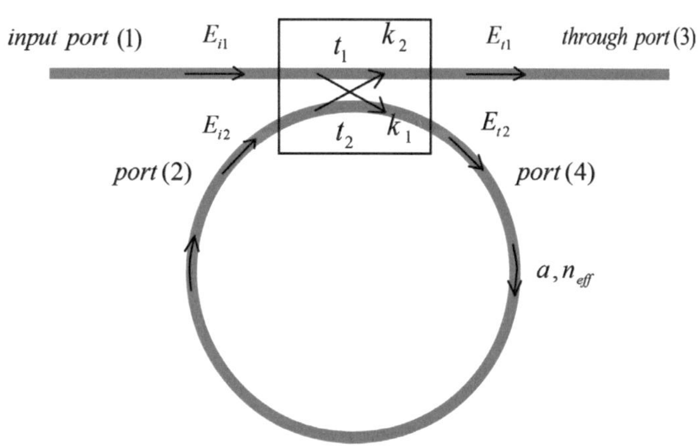

Figure 2.
All-pass fiber-optic ring resonator configuration ($\kappa_1 \approx \kappa_2 = \kappa$ and $t_1 \approx t_2 = t$).

where sign and i-factor in $i\kappa$ elements of first matrix on the right side of equation are due to cross-coupling phase shift caused by coupler. Also, since this matrix is an unitary matrix, we find that accomplishes the following relationship between bar- (t) and cross- (κ) coupling coefficients (assuming both are independent of frequency):

$$t^2 + \kappa^2 = 1 \tag{5}$$

The feedback path of length $2\,\pi\,R$ (R being the ring radius) connects the output from port 4 back into input port 2, where the field can be expressed as:

$$E_{i2} = e^{-\alpha 2\pi R} e^{i\beta 2\,\pi R} E_{t2} \equiv a e^{i\phi} E_{t2} \tag{6}$$

where α is the attenuation factor of the fiber, a-factor represents the ring single-pass field amplitude transmission (taking into account the fiber attenuation), $\beta = (2\pi/\lambda) n_{eff}$ is the propagation constant of propagating mode in the fiber (where in time, n_{eff} is the effective index of the fiber) and ϕ represents the single-pass ring round-trip phase shift.

Since that adding or subtracting an integer number m of $2\,\pi$ radians from single-pass phase shift does not change the value of the function, the single-pass phase shift for all resonances is defined such that its value is zero for a local resonance of interest. Furthermore, because the single-pass phase shift is directly related to the radian frequency as $\phi = \omega T_R$, where T_R is the transit time of the round-trip of ring resonator and then ϕ is clearly representative of a *normalized frequency detuning*. Following the previous derivations, one can demonstrate that the transmission power P_{t1} in the output port (3) of the straight waveguide can be expressed as:

$$P_{t1} = |E_{t1}|^2 = \frac{a^2 + |t|^2 - 2a|t|\cos(\phi + \phi_c)}{1 + a^2|t|^2 - 2a|t|\cos(\phi + \phi_c)} \tag{7}$$

where $t = |t|\exp(j\phi_c)$, $|t|$ representing the coupling losses and ϕ_c represents the phase of the coupler. Making a similar deduction, we can reach the circulating power in the ring, which is:

$$P_{i2} = |E_{i2}|^2 = \frac{a^2\left(1 - |t|^2\right)}{1 + a^2|t|^2 - 2a|t|\cos(\phi + \phi_c)} \tag{8}$$

On resonance, condition $(\phi + \phi_c) = 2\pi m$ is fulfilled, where m is an integer, and then Eqs. (7) and (8) becomes:

$$P_{t1} = |E_{t1}|^2 = \frac{(a - |t|)^2}{(1 - a|t|)^2} \tag{9}$$

$$P_{i2} = |E_{i2}|^2 = \frac{a^2\left(1 - |t|^2\right)}{(1 - a|t|)^2} \tag{10}$$

where, as we have said above, a is the loss coefficient of the ring (for zero loss $\alpha = 0$, then $a = 1$). The special case that occurs when $a = |t|$ (or $\kappa^2 = \alpha L$) is named *critical coupling* and is of high interest in applications. This happens when internal losses around the ring are equal to the coupling losses.

Then, the transmitted power to through port (3) becomes 0. The critical coupling is due to destructive interference between the direct beam and the cross-coupled beam of light, both after passing their respective coupling zone. A condition termed *undercoupling* does occur when the ring loss exceeds the coupling strength ($t > a$ or equally: $t^2 > \alpha L$, also $\kappa^2 < \alpha L$). An *overcoupling* condition takes place, when the round-trip loss does not exceed the coupling strength ($t < a$ or equally: $t^2 < \alpha L$, also $\kappa^2 > \alpha L$) and is the conventional mode of operation for an all-pass resonator. **Figure 3**(**a**) and (**b**) displays the effect of attenuation on the transmission and build-up factor for the overcoupled, critically coupled and undercoupled regimes. Experimental demonstration of the transmission characteristics in each of these regimes for a fiber all-pass filter ring resonator can be found in [22].

A good idea of the behavior of the system, when resonance condition accomplishes, can be obtained when one represents the T *transmission function* defined as the ratio P_{t1}/P_{i1} (i.e., the relationship between the through-port power and input-port power) versus normalized detuning ϕ, **Figure 3**(a). Starting from input field expressions and after some derivations, the following equation can be found:

$$\mathcal{T} = \frac{P_{t1}}{P_{i1}} = \left|\frac{E_{t1}}{E_{i1}}\right|^2 = \frac{a^2 - 2|t|a\cos\phi + |t|^2}{1 - 2|t|a\cos\phi + |t|^2 a^2} \tag{11}$$

Figure 3(**a**) represents the T transmission function for an all-pass ring resonator with $t = 0.75$ and varying loss a as a parameter (from $a = 0.00$ up to $a = 0.99$). It can be seen that, in this case, critical coupling occurs when $a = t = 0.75$; therefore, the T transmission function drops to zero when $\phi = 0$ for this a-factor value. As a

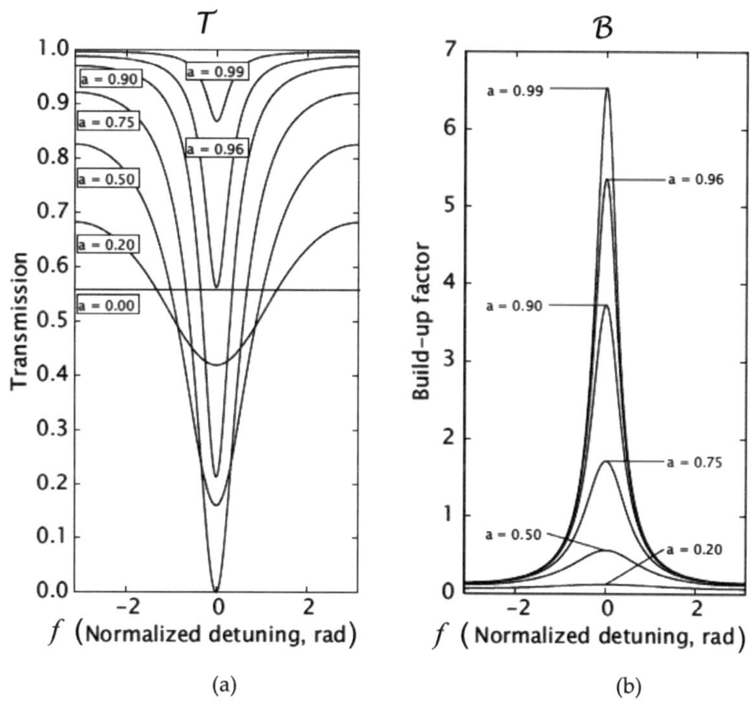

Figure 3.
T transmission function (a) and B build-up factor (b) of ring-resonator, both as a function of normalized detuning ϕ, (curves extracted from [21], with permission).

increases (up to 1) or decreases (to 0.00) from 0.75, the corresponding transmission curve arises its minimum point from 0. For a = 0.00, T transmission function reaches a constant value of 0.75^2 = 0.5625 for all values of ϕ. Another function that allows to obtain a good idea of the behavior of the system is the so-called B *intensity build-up factor* as a function of ϕ (normalized detuning), defined as:

$$\mathcal{B} = \frac{P_{i2}}{P_{i1}} = \left|\frac{E_{i2}}{E_{i1}}\right|^2 = \frac{\left(1-|t|^2\right)a^2}{1-2|t|a\cos\phi+|t|^2a^2} \Rightarrow \lim_{\phi=2\pi m,\, a=1} \frac{\left(1-|t|^2\right)a^2}{1-2|t|a\cos\phi+|t|^2a^2} = \frac{1+|t|}{1-|t|} \approx \frac{4}{\kappa^2} \tag{12}$$

B function can be thought as the ratio of the circulating light intensity inside the ring to incident light intensity. **Figure 3(b)** represents the B function (or intensity build-up factor) versus normalized detuning ϕ for the same all-pass ring than the one in **Figure 3(a)**, when varying loss a is taken as a parameter (from a = 0.20 up to a = 0.99). The last result in Eq. (12) refers to case in which incident light is resonant with the ring ($\phi = 2m\pi$) and attenuation is negligible (a = 1). This happens when constructive interference is produced at the coupler port (4) between the coupled light entering the ring coming from waveguide and the recirculation light coming from input port (3). This way, the circulating intensity of light around the ring results reinforced. However, for this constructive interference to take place, a coherent source is essential at input port (1). A passive ring resonator under these conditions attains the maximum ratio of circulating power to incident power that can be achieved. **Figure 4** shows two successive resonances versus normalized detuning with peak values given by right side of Eq. (12).

The spectral characteristics of the all-pass fiber-ring resonator configuration (one waveguide, one resonator, as represented in **Figure 2**) as optical filter can better be viewed when spectral response of through port versus vacuum wavelength is represented. A notch type filter is obtained from observation of this resonance curve (**Figure 5**). For cross-coupling value of 10% ($\kappa^2 = 0.10$), the optical intensity in the ring can be 40 times higher than the intensity in the input port of the waveguide. This way, since the intensity in the fiber ring can be much higher

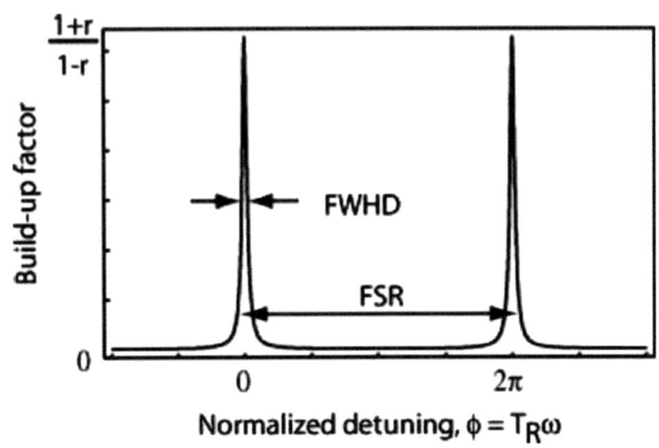

Figure 4.
Build-up factor versus normalized detuning for all-pass fiber ring resonator (from [21], with permission).

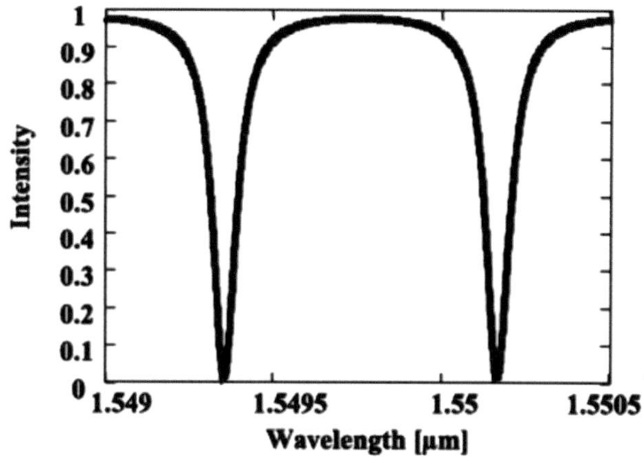

Figure 5.
Spectral response of through port of an all-pass fiber ring resonator (notch-type filter).

than in the bus, fiber ring resonators can be used for nonlinear optics applications with low input intensities.

Ring resonator filters can be characterized by four parameters, which are generally used to describe their performance, namely, *the free spectral range* (FSR), the *full width at half maximum* (FWHM), *the finesse* F and *the quality factor* Q. The FSR $\Delta\lambda$ is the distance between two successive resonance peaks (difference between the wavelengths corresponding to two successive resonant conditions). When neglecting the wavelength dependence of the fiber-optic effective refractive index n_{eff}, approximate calculation of FSR $\Delta\lambda$ is very simple [20], as follows:

$$FSR = \Delta\lambda = -\frac{2\pi}{L}\left(\frac{\partial\beta}{\partial\beta}\right)^{-1} \approx \frac{\lambda^2}{n_{eff}L} \tag{13}$$

This approximation is derived from the resonant condition next to a resonance found for the used propagation constant. A more accurate calculation is obtained taking into account the chromatic dispersion of the fiber. Then, the group refractive index of the fiber can be used instead of the effective index, so that the new expression for FRS to be:

$$FSR = \Delta\lambda = \frac{\lambda^2}{n_g L} \tag{14}$$

The second parameter in importance for describing the performance of the optical filter is the FWHM. It is defined as the full spectral width at half maximum (or 3 dB bandwidth $2\delta\lambda$). In the case of all-pass filter, FWHM must be redefined as full width at half depth (FWHD) and its value can be calculated as follows (in absence of ring internal losses, that is, $a = 1$):

$$FWHD = 2\delta\lambda = \frac{\lambda^2}{2\pi^2 n_g R}\frac{1-|t|}{\sqrt{|t|}} \xrightarrow{|t|\approx 1} \approx \frac{\lambda^2\kappa^2}{4\pi^2 n_g R} \tag{15}$$

Third parameter is the finesse F. The finesse is a figure of merit of the filter that can be defined as the ratio between FSR and FWHD. So that implementing this definition and in absence of ring internal losses, the resulting expression for the finesse of an all-pass fiber resonator is [23]:

$$\mathcal{F} = \frac{FSR}{FWHD} \approx \frac{\frac{\lambda^2}{2\pi n_g R}}{\frac{\lambda^2 \kappa^2}{4\pi^2 n_g R}} = \frac{2\pi}{\kappa^2} \tag{16}$$

Finally, the fourth parameter is the quality factor Q. It represents a measure of the sharpness of the resonance. It can be defined as the ratio of the operation wavelength and the resonance width (FWHD). Its calculation is as follows:

$$\mathcal{Q} = \frac{\lambda}{FWHD} \approx \frac{\lambda}{\frac{\lambda^2 \kappa^2}{4\pi^2 n_g R}} = \frac{\lambda}{\frac{\lambda^2 \kappa^2}{2\pi n_g L}} = \frac{2\pi n_g L}{\lambda \kappa^2} = \frac{n_g L}{\lambda} \mathcal{F} \tag{17}$$

The quality factor Q can also be regarded as the total energy stored in the ring divided by the energy lost per optical cycle.

3. Analysis of the fiber-optic add-drop filter ring resonator configuration

The basic fiber-optic add-drop filter consists of one input-waveguide, one output-waveguide and one fiber-ring resonator such as the one represented in **Figure 6**. Both of the waveguides are side-coupled to ring resonator. The four ports of structure are defined as input port, throughput port, drop port and add port. Viewed from drop port acting as the optical output, this configuration behaves as a narrow-band amplitude filter that can add or drop a spectral band from an optical input. This configuration is mathematically equivalent to a well-known Fabry-Pérot interferometer. Therefore, the equations for transmission coefficients are easy to obtain. For simplification, E_{i1} is defined with unitary rms amplitude. Then, the throughput E_{t1} mode rms amplitude in the input-waveguide is given by:

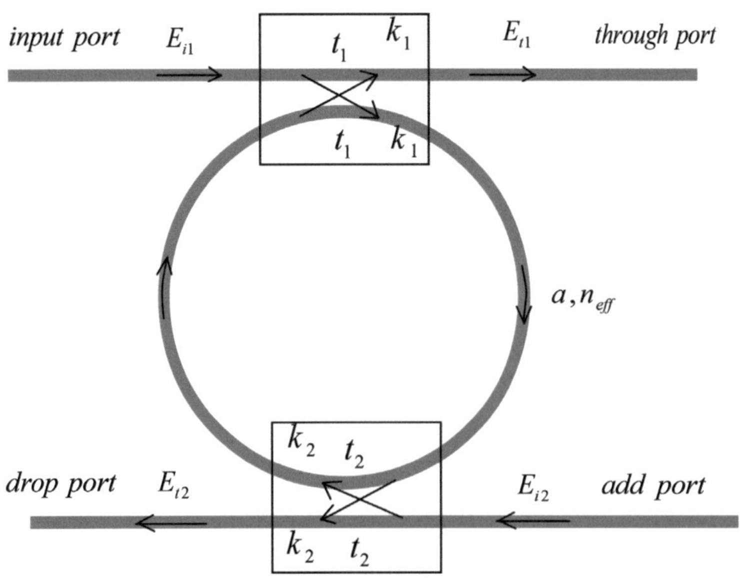

Figure 6.
Add-drop fiber-optic ring resonator configuration (for symmetrical coupling: $\kappa_1 \approx \kappa_2 = \kappa$ and $t_1 \approx t_2 = t$).

$$E_{t1} = \left[t_1 + \frac{t_2 \kappa_1 \kappa_2 a_{1/2}^2 e^{j\theta\left(|t_1|^2+|\kappa_1|^2\right)}}{1 - t_1 t_2 a_{1/2}^2 e^{j\theta}} \right] E_{i1} = \frac{t_1 - t_2 a e^{j\theta}}{1 - t_1 t_2 a e^{j\theta}} \tag{18}$$

In this equation, $a_{1/2}$ and $\theta_{1/2}$ are used, which are the half round-trip loss and phase, respectively. It is $a = a_{1/2}^2$ and $\theta = 2\theta_{1/2}$. The mode amplitude in the ring has to pass the second coupler to become the new dropped mode rms amplitude E_{t2}. Its rms amplitude value is given by:

$$E_{t2} = \frac{\kappa_1 \kappa_2 a_{1/2} e^{j\theta_{1/2}}}{1 - t_1 t_2 a e^{j\theta}} \tag{19}$$

At resonance condition ($\theta = m2\pi$, m integer), the output intensity from the drop port is given by:

$$P_{t2-resonance} = |E_{t2-resonance}|^2 = \frac{\left(1-t_1^2\right)\left(1-t_2^2\right)a}{\left(1 - t_1 t_2 a\right)^2} \tag{20}$$

The through port mode amplitude E_{t1} will be zero at resonance for identical symmetrical couplers ($t_1 = t_2$), when $a = 1$ (zero loss). This means that on resonance, all input optical power will be coupled to resonator and could be extracted by drop port. The value $a = 1$ (zero loss) can only be achieved by a given value of gain incorporated in the resonator ring to compensate the waveguide losses. In a pure passive ring resonator with defined waveguide constituents, the loss coefficient a is constant. By adjusting coupling parameters t_1 and t_2 to the loss coefficient a, minimum intensity can be achieved (ideally, $P_{t1} \approx 0$). From Eq. (18), one can obtain:

$$a = \frac{t_1}{t_2} \tag{21}$$

According with Eq. (18), the transfer function of through port (relative to input port) is given by:

$$T_{through} = \left|\frac{E_{t1}}{E_{i1}}\right|^2 = \frac{t_2^2 a^2 - 2t_1 t_2 a \cos\phi + t_1^2}{1 - 2t_1 t_2 a \cos\phi + \left(t_1 t_2 a\right)^2} \xrightarrow[a=1,\, t_1=t_2=t]{} \frac{\frac{4t^2}{(1-t^2)^2}\sin^2\left(\frac{\phi}{2}\right)}{1 + \frac{4t^2}{(1-t^2)^2}\sin^2\left(\frac{\phi}{2}\right)} \tag{22}$$

Taking into account, Eq. (19), the transfer function of drop port (relative to input port), will be:

$$T_{drop} = \left|\frac{E_{t2}}{E_{i1}}\right|^2 = \frac{\left(1-t_1^2\right)\left(1-t_2^2\right)a}{1 - 2t_1 t_2 a \cos\phi + \left(t_1 t_2 a\right)^2} \xrightarrow[a=1,\, t_1=t_2=t]{} \frac{1}{1 + \frac{4t^2}{(1-t^2)^2}\sin^2\left(\frac{\phi}{2}\right)} \tag{23}$$

This way, the transmission spectral response of through port corresponds to an approximately Lorentzian curve centered on a resonance, where a narrow frequency band has been extracted (see **Figure** 7). These extracted band do appear (in inverted shape) in the spectrum response of drop port, as it can also be seen in same figure. In the case of add-drop filter, if attenuation is negligible ($a \approx 1$), *critical coupling* occurs at symmetric coupling ($\kappa_1 = \kappa_2 = \kappa$). For non-negligible single-pass resonator loss, critical coupling occurs when single-pass ring losses match the

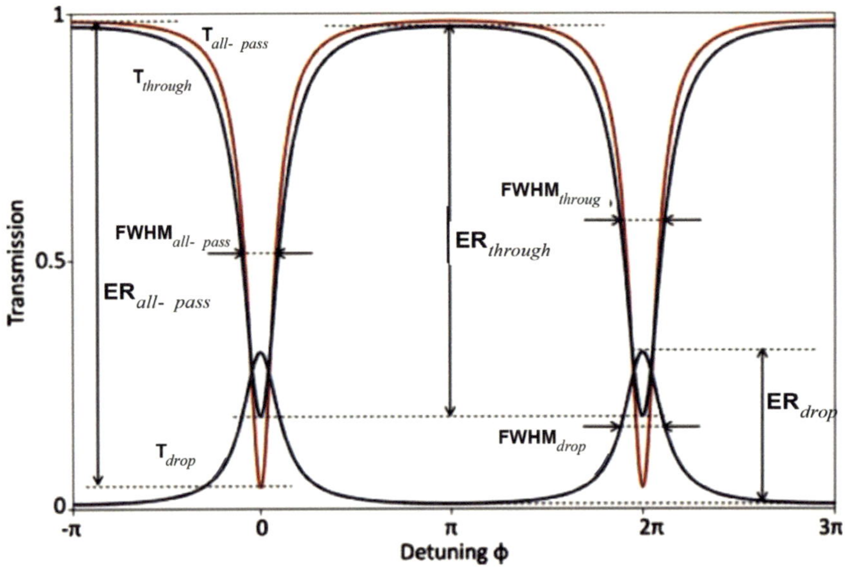

Figure 7.
Spectral response of a fiber-optic add-drop filter (adapted from [24], with permission).

coupling, so it verified the following condition: $t_2 a = t_1$. Transfer functions, FWHM and extinction ratio (ER) relative to all-pass filter, through and dropt ports of add-drop filter are plotted.

It is worth to make an interpretation and comparison of the add-drop filter configuration with respect to the all-pass filter one. In the upper coupler (**Figure 6**), interference takes place between the coupled input field and ring circulating field. However, the lower coupler does not experience any interference (provided that excitation is from the upper waveguide only). The lower coupler may thus be viewed simply as a tap waveguide that leaks power out of the ring into the drop port. For an add-drop filter, it is desirable to operate at critical coupling for achieving the complete extinguishment of a frequency band in the through port. In this case, the sum of all losses incurred in the resonator, even the loss at the out-coupled drop port, must be taken into account. Calculations of FSR, FWHM, B build-up factor, finesse F and quality factor Q are given below according to the results found in Refs. [21, 22].

$$FSR = \Delta\lambda = \frac{\lambda^2}{n_{eff} L} \tag{24}$$

$$FWHD \approx \frac{\lambda^2 \kappa^2}{\pi L n_{eff}} \tag{25}$$

$$\mathcal{F} = \frac{FSR}{FWHD} \approx \frac{\pi}{\kappa^2} \tag{26}$$

$$\mathcal{Q} = \frac{\lambda}{FWHD} \approx \frac{\pi n_g L}{\lambda \kappa^2} = \frac{n_g L}{\lambda} \mathcal{F} \tag{27}$$

where the approximations again apply to symmetrically coupled, low-loss and weakly coupled microrings. Note that, comparing Eqs. (27) and (17), with the same optical parameters, add-drop filter has the half value of Q-factor (and also the finesse F) than that the all-pass one. This result could be expected since an all-pass filter configuration has half the loss due to external coupling than an add-drop one.

4. Numerical simulation of fiber-optic all-pass filter ring resonator configuration

As announced at the beginning of this chapter, this section will deal specifically with the numerical simulation of a typical FOPRRI in all-pass filter configuration. First, effective index method (EIM) [25, 26] will be applied to this structure in which main constituents are made of a standard single-mode optical fiber (coupler and ring resonator). Secondly, coupled mode theory (CMT) [27–33] will be performed to the same structure. In third time, finite element method (FEM) [34] is performed. Finally, 3D-FDTD (finite difference time domain) [35, 36] was applied to same structure. The results of all these numerical simulations are compared to each other to see which one best match each specific application.

Optical layout scheme for EIM and CMT numerical simulation is represented in **Figure 8**. OptSim® Circuit tool (from Synopsys™ Software) was applied to obtain the transmission power spectrum of this resonant structure, which takes the effective index method (EIM) as a numerical simulation algorithm. A side-polished fiber-optic 4 × 4 directional coupler is used to couple the waveguide fiber to fiber ring resonator (here implemented by a series of two waveguides, 1 and 2, respectively). A standard SMF-28 (Corning™ Glass) single-mode optical fiber is assumed for both, waveguide and ring resonator as constituent materials. Optical parameters of this fiber are: basis material = fused-silica (core = boron-silicate, cladding = silica), step refractive index, core diameter = 8.2 μm, cladding diameter = 125 μm, effective refractive indexes: n_{core} = 1.467, $n_{cladding}$ = 1.460, numerical aperture (NA) = 0.143, maximum attenuation = 0.35 dB/km (at 1310 nm wavelength) and h-parameter = $2 \times 10^{-6}\ m^{-1}$. Gap between waveguide and ring resonator is 200 nm. Normalized Gaussian profile continuous wave (CW) optical field of 1310 nm vacuum wavelength was launched to the system from a wideband tunable laser source. To obtain the pass-through port power transmission spectrum, wavelength sweeping of laser source was made in order to fix each wavelength value for its respective measured optical power. The computation is iterated 30 times to obtain the steady-state solution. Obtained transmission power spectrum is represented in **Figure 9**.

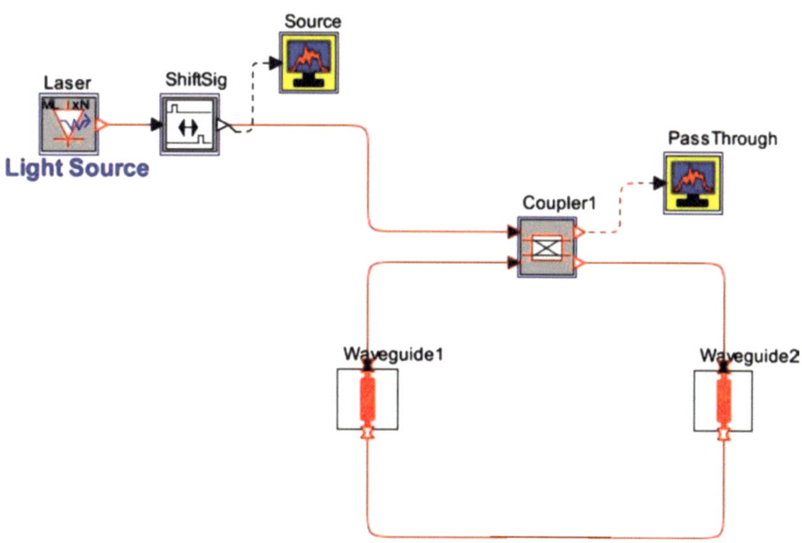

Figure 8.
All-pass FOPRRI configuration: optical layout scheme for EIM and CMT numerical simulations.

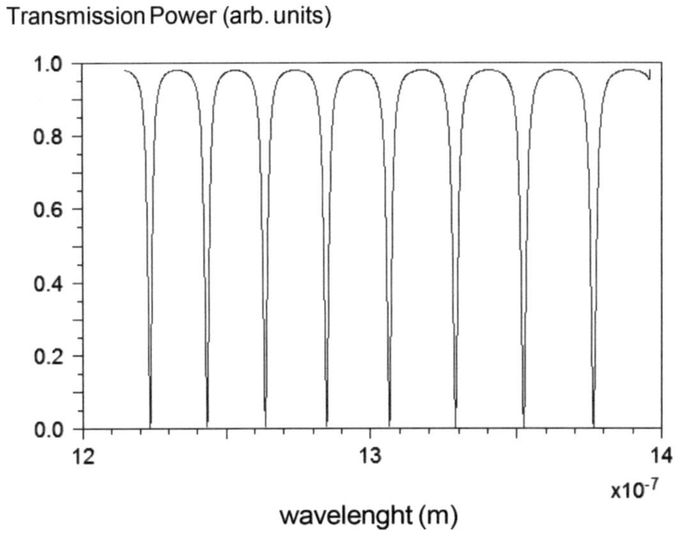

Figure 9.
All-pass FOPRRI configuration: transmission power spectrum.

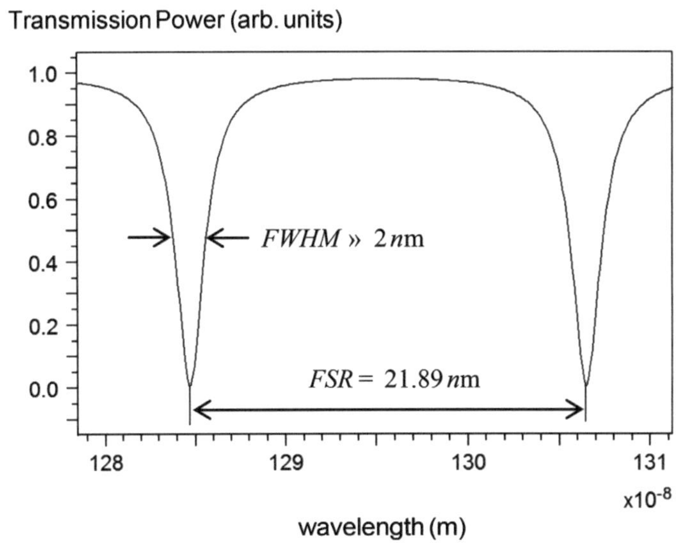

Figure 10.
All-pass FOPRRI configuration: expanded transmission power spectrum around 1295 nm.

As it can be seen on expanded transmission spectrum showed in **Figure 10**, a resonance condition is achieved for 1285 nm wavelength. Since the effective refractive index of ring resonator is 1.467 and its optical path length is 51.420 μm, a FSR of 21.89 nm is obtained as it can also be observed in this same figure. This value agrees accurately with theoretical value as calculated from Eq. (13).

On the other hand, measured value of FWHM is approximately 2 nm. This value is also in frank agreement with value calculated analytically by means of Eq. (15), taking into account 8.184 μm as medium ring resonator radius. Then, a calculation from Eq. (16) gives the value of 10.95 for F finesse. As a consequence of last, calculation from Eq. (17) yields 642.80 for the Q quality-factor of the system.

Coupled mode theory (CMT), which uses BPM method, was applied to this structure by means of BeamPROP® tool (Synopsys™ Software) [32]. This method

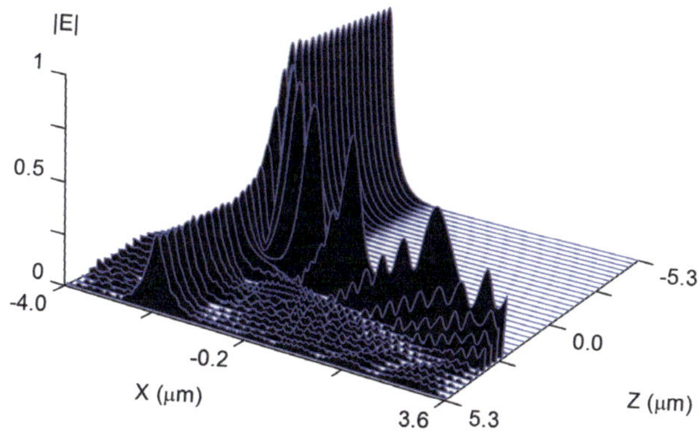

Figure 11.
CMT method numerical simulation result (from BeamPROP® tool) for all-pass FOPRRI configuration (resonance at 1285 nm).

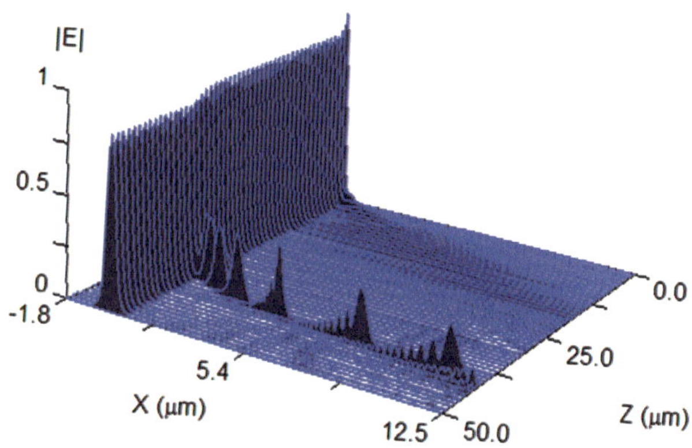

Figure 12.
FEM numerical simulation (from FemSIM® tool) for all-pass FOPRRI configuration (resonance at 1285 nm).

also takes into account the effective index of the ring resonator material, so the expected results from it will be improved compared with those obtained with the EIM method alone. Further in-depth readings on BPM method can be found in Refs. [37–42]. **Figure 11** shows the result of numerical simulation of all-pass ring resonator considered here.

Finite element method (FEM) was applied to this structure by means of FemSIM® tool (Synopsys™ Software) [34]. This method also takes into account the effective index of the ring resonator material, so the expected results from it will be improved compared with those obtained with the EIM method alone. Additional treatment on the FEM method can be found in the Refs. [43–47]. The result of FemSIM® of all-pass ring resonator analyzed here is represented in **Figure 12**.

FullWAVE® (Synopsys™ Software) [36] is a full-vectorial numerical computation tool that allows the simulation of light wave propagation in an arbitrary geometry system. The simulation engine is based on the finite difference time domain (FDTD) method. The FDTD algorithm implements rigorous and complete time-space solution to Maxwell's equations. This method offers an efficient field wave propagation solution in a photonic-integrated structure when other methods such as beam propagation method (BPM) or finite element method (FEM) cannot cope with the structure geometry or do not provide a precise solution. Additional treatment on the FDTD method can be found in Refs. [48–52]. **Figure 13** represents the result of FDTD numerical simulation of all-pass ring resonator studied here.

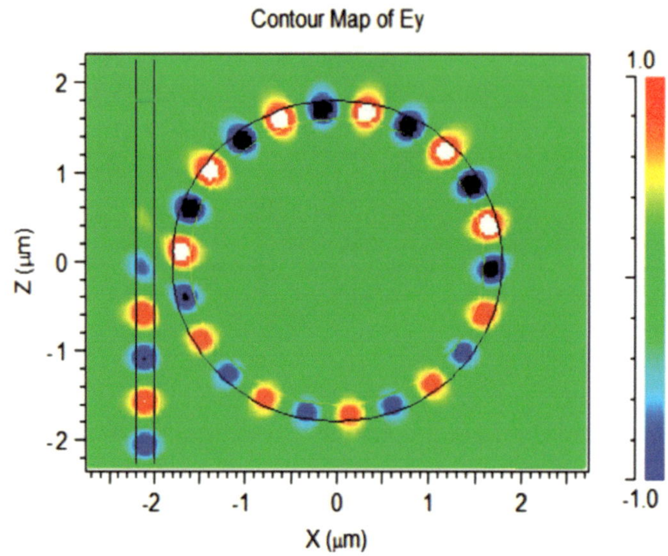

Figure 13.
2D-FTDT numerical simulation results for an all-pass FOPRRI (resonance at 1285 nm).

Comparison	EIM	CMT (BPM)	FEM	FDTD
All-pass filter	Good	Very good	Outstanding	Excellent
Add-drop filter	Good	Very good	Outstanding	Excellent

Table 1.
Comparison of numerical simulation methods (EIM, CMT, FEM and FDTD) for all-pass/add-drop filters.

Table 1 shows a comparison of the results of the four methods used (EIM, CMT, FEM and FDTD) for all-pass filter and add-drop filter configurations, taking into account their suitability for accuracy and calculation time. In view of transmission spectrum of the all-pass filter configuration, this can find applications such as optical band-pass filter, optical delay line, optical switch and active ring modulator.

5. Numerical simulation of fiber-optic add-drop filter ring resonator configuration

As realized on all-pass configuration, this section will be dedicated to numerical simulation of a typical FOPRRI in add-drop filter configuration. Then, successive numerical simulations by applying effective index method (EIM), coupled mode theory (CMT), finite element method (FEM) and 3D finite difference time domain (FDTD) will be performed to this configuration. Finally, the comparison of all these numerical simulations will be carried out to verify, which is the most appropriate for each type of application.

Optical layout scheme for EIM and CMT numerical simulation is represented in **Figure 14**. OptSim® Circuit tool (from Synopsys™ Software) with the effective index method (EIM) as computational algorithm was applied here to obtain the transmission power spectrum of resonant structure. The configuration analyzed can be obtained from the one considered in Section 4 by adding a second fiber 4 × 4 directional coupler and a second side-coupled fiber waveguide. Also, standard SMF-28 (Corning™ Glass) single-mode optical fiber is assumed for both, upper and

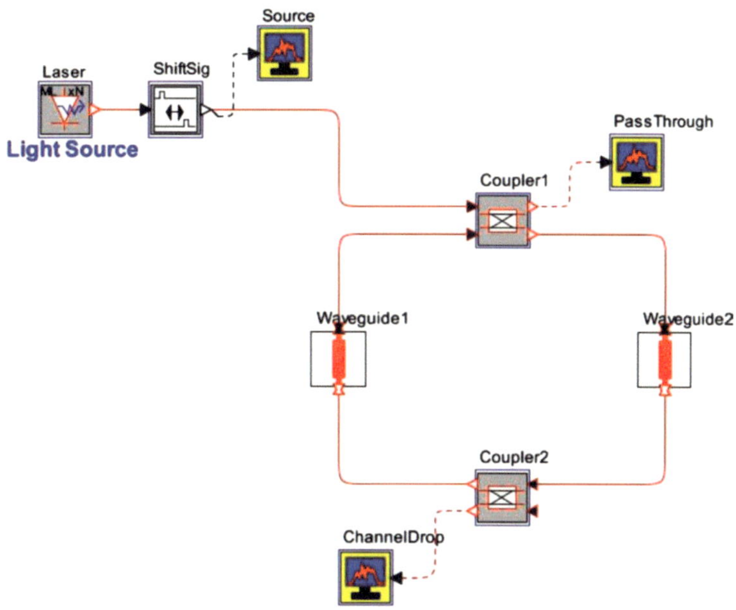

Figure 14.
Add-drop FOPRRI configuration: optical layout scheme for EIM and CMT numerical simulations.

lower waveguides and ring resonator as constituent material. Gap between waveguide and ring resonator is 200 nm. Normalized Gaussian profile continuous wave (CW) optical field of 1310 nm vacuum wavelength was launched to the system from a wideband tunable laser source. To obtain the power transmission spectra of pass-through and drop ports, wavelength sweeping of laser source was made in order to fix each wavelength value for its respective measured optical power. The computation is iterated 30 times to obtain the steady-state solution. The resulting transmission power spectra are depicted in **Figure 15(a)**–(**c**), respectively. As it can be seen on expanded drop-port transmission spectrum showed in **Figure 10**(**c**), a resonance condition is achieved for 1285 nm wavelength. Since the effective refractive index of ring resonator is 1.467 and its optical path length is 51.420 μm, a FSR of 21.89 nm is obtained as it can also be observed in this same figure. This value agrees accurately with theoretical value as calculated from Eq. (24). On the other hand, measured value of FWHM is approximately 4 nm. This value is also in frank agreement with value calculated analytically by means of Eq. (25).

Then, a calculation from Eq. (26) gives the value of 5.475 for F finesse. As a consequence of last, calculation from Eq. (27) yields 321.40 for the Q quality-factor of the system. These results are the expected, taking into account that for the same coupling coefficients, waveguide and ring resonator geometry, the values F and Q for add-drop configuration must be half than those for the all-pass ones.

Comments about fundamentals and particularities of BPM, FEM and FDTD methods are given in the following. For an optical waveguide as cylindrical optical fiber, where light travels almost completely confined to its core medium, propagation mechanism can be simply described by means of the four Maxwell's equations. Then, BPM, FEM and FDTD methods assume that the solution to Maxwell's equations follows the slowly varying envelope approximation. This means that the field's amplitude variation is very small as compared to the variation in the phase and propagation distance. BPM is a fast method, which is applicable to a system whose physical structure and properties vary slowly with the geometry. The FDTD method uses the standard finite difference method to solve the Maxwell's equations. It is also

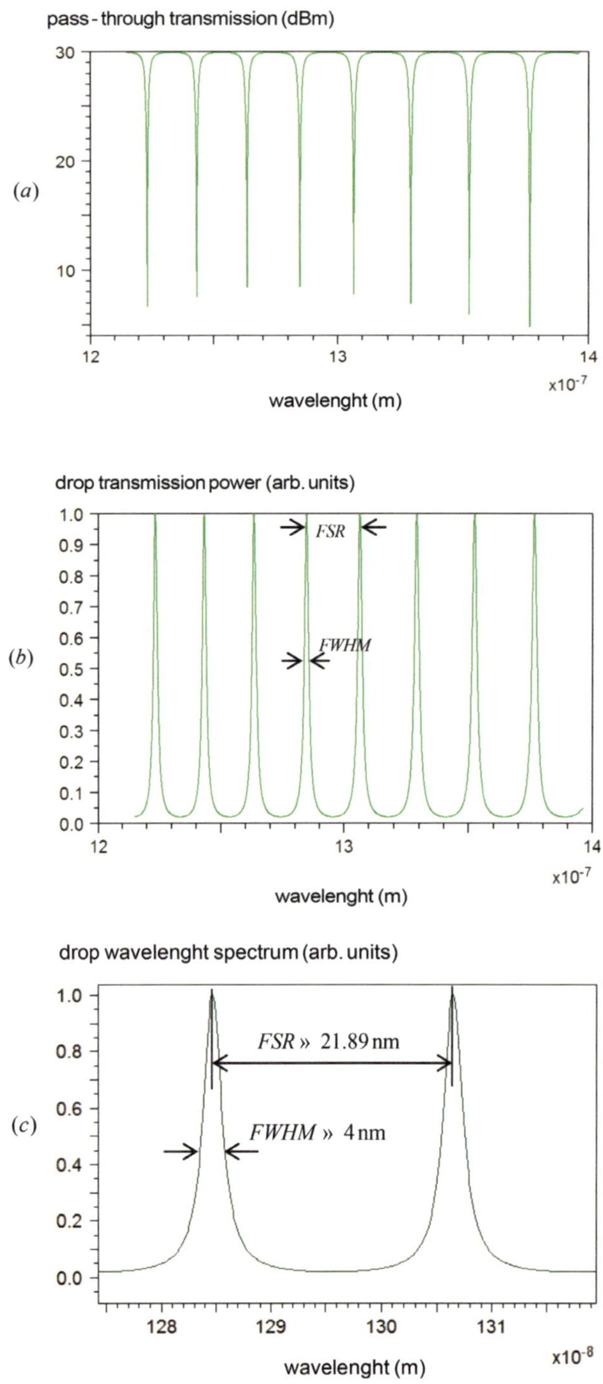

Figure 15.
Transmission spectra of add-drop configuration: (a) pass-through port, (b) drop port and (c) drop-port (expanded).

known as the full wave method. It discretizes the equations and solves them using a linear algebraic algorithm.

To accelerate the finite difference method simulation, standard approaches such as Krylov subspace-based model order reduction or multi-grid analysis are used. It can be used for all types of structures. As compared with BPM, FDTD is at least an order of magnitude slower. The basic elements that are of application in the case of

FDTD method allow as broad as the Maxwell's equations they embody in discretized form, the basic formulation using centered finite difference expressions, which can treat virtually any material type and geometry and the basic resources needed to implement the technique. The FDTD method treats transients in the time domain (e.g., pulses), and it is applicable over the computationally unpredictable resonance region in which a wavelength is comparable with to the object interaction size. The FDTD method offers many advantages as an electromagnetic modeling, simulation and analysis tool. Its capabilities include: 1) broadband response predictions centered about the system resonances, 2) arbitrary 2-D or 3-D model geometries, 3) interaction with an object of any conductivity from that of a perfect conductor, to that of a real metal, to that of low or zero conductivity, 4) frequency-dependent constitutive parameters for modeling most materials (lossy dielectrics, magnetic materials, anisotropic plasmas and magnetized ferrites) and 5) any type of response, including far fields derived from near fields, such as scattered fields, antenna patterns, radar-cross sections, surface response, current and power density and penetration/coupling. FTDT is an appropriate technique for time-domain analysis of passive microwave and millimeter wave structures. The strength of FDTD method lies in its capability to model any volumetric structure with what are typically rectangular cells. No requirements for symmetry or smooth surfaces are needed. Also, FDTD has been successfully used in modeling micro-structured micro-strip structures.

Coupled mode theory (CMT), which uses BPM method, was applied to this structure by means of BeamPROP® tool (Synopsys™ Software). This method also takes into account the effective index of the ring resonator material, so the expected results from it will be improved compared with those obtained with the EIM method alone. BPM numerical simulation results can be seen in **Figure 16**.

Finite element method (FEM) was applied to this structure by means of FemSIM® tool (Synopsys™ Software). This method also takes into account the effective index of the ring resonator material, so the expected results from it will be improved compared with those obtained with the EIM method alone. FEM numerical simulation of this case is represented in **Figure 17**.

Result of a FDTD method numerical simulation applied to the add-drop filter configuration considered here is represented on **Figure 18** (resonance takes place at 1285 nm wavelength). As it was said above, for this structure Eq. (25) gives the value of 5.475 for F finesse and Eq. (26) yields 321.40 for the Q quality-factor of the system.

As a consequence of transmission spectrum of add-drop filter configuration, this can find applications such as optical band-pass filter, optical delay line,

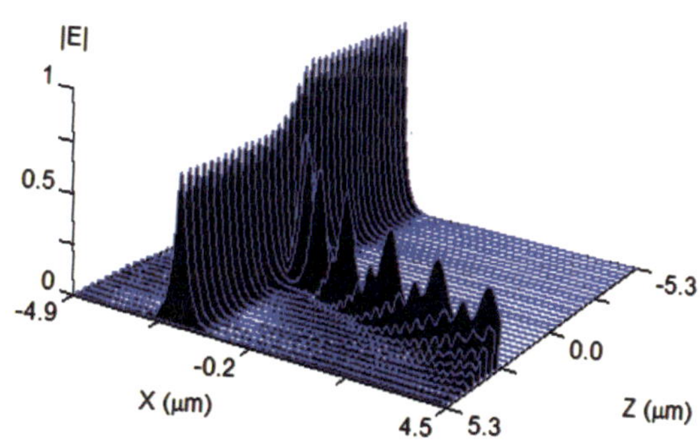

Figure 16.
CMT method numerical simulation result (from BeamPROP® tool) for all-pass FOPRRI configuration (resonance at 1285 nm).

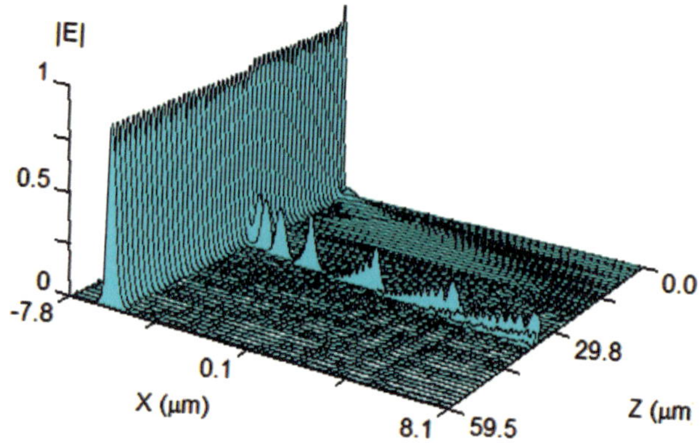

Figure 17.
FEM numerical simulation (from FemSIM® tool) for all-pass FOPRRI configuration (resonance at 1285 nm).

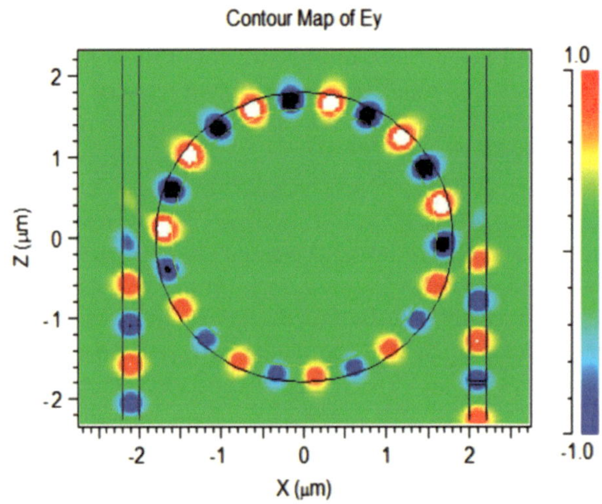

Figure 18.
2D-FTDT numerical simulation results for an add-drop FOPRRI (resonance at 1285 nm).

optical switch, inertial rotation sensor, photonic biosensor, tailored response filter and active ring modulator.

6. Performance comparison of numerical simulation methods used

Table 2 shows performance comparison of all four methods used (EIM, CMT, FEM and FDTD) for both the configurations: all-pass filter and add-drop filter and

Criterion/Method	EIM	CMT (BPM)	FEM	2D-FDTD/3D-FDTD
Computational time (sec) 2D	< 0.05	<0.10	1.7	4.3
Computational time (sec) 3D	68	100	547	1170
Memory (MByte)	16	17	17	17
Error (%)	<7	<1	<1	<1

Table 2.
Comparison of computational time, memory resources and error methods (EIM, CMT, FEM and FDTD).

taking into account three criteria: spent computational time (seconds), memory resources applied (MByte) and error found (%).

7. Conclusions

A conventional model of a fiber optic passive ring resonator interferometer (FOPRRI) was theoretically analyzed in two main configurations: all-pass filter and add-drop filter. Then, four numerical simulation methods (EIM, CMT, FEM and 3D-FDTD) are successively applied on each of these two configurations. In both cases, results showed that EIM method is the most suitable when no very accuracy of results is needed (error < 7%) and low time of computation is desirable (<70 s). However, when high accuracy is considered the most important issue, 3D-FDTD is the best method (error < 1%) at the expense of a much larger computing time (≈1200 sec) than others. CMT (BPM) method provides substantial performance results with computational time similar than that EIM one, but an error < 1%. Finally, FEM is suitable for simulations that can be approximated with 2D techniques since it provides substantial performance results (error < 1%) with computational time only an order of magnitude above EIM or CMT.

Acknowledgements

The author wishes to thank the Integrated Optics and Optoelectronics Laboratory of Faculty of Sciences of University of Oviedo-Spain for technical support and assistance provided in the realization of this work.

Author details

Ramón José Pérez Menéndez
UNED, Lugo, Spain

*Address all correspondence to: ramonjose.perez@lugo.uned.es

References

[1] Hariharan P. Optical Interferometry. 2nd ed. London: Elsevier, Academic Press; 2003

[2] Sharma U, Wei X. Fiber optic interferometric devices. Chapter 2. In: Kang JU, editor. Fiber Optic Sensing and Imaging. New York: Springer; 2013. pp. 29-53

[3] Yin S, Ruffin PB, Yu FTS, editors. Fiber Optic Sensors. 2nd ed. Boca Raton FL (USA): CRC Press; 2008. pp. 333-366

[4] Yariv A, Yeh P. Optical Resonators from Photonics. Chapter 4. 6th ed. Oxford University Press, New York; 2007. pp. 156-210

[5] Vali V, Shorthill RW. Fiber ring interferometer. Applied Optics. 1976; **15**(5):1099-2000

[6] Vali V, Shorthill RW. Fiber laser gyroscopes. Proceedings of SPIE. 1976; **76**:110

[7] Vali V, Shorthill RW. Ring interferometer 950 m long. Applied Optics. 1977;**16**:290

[8] Ezequiel S, Balsamo SR. Passive ring resonator laser gyroscope. Applied Physics Letters. 1977;**30**:478

[9] Meyer RE, Ezequiel S. Passive fiber-optic ring resonator for rotation sensing. Optics Letters. 1984;**8**:644-646

[10] Rao YJ, Jackson DA. Principles of fiber-optic interferometry. In: Grattan LS, Meggitt BT, editors. Optical Fiber Sensor Technology. Boston, MA: Springer; 2000. pp. 167-191

[11] Yve CY, Peng JD, Liao YB, Zhou BK. Fibre ring resonator with finesse of 1260. Electronics Letters. 1988;**24**:622-623

[12] Stokes LF, Chodorow M, Shaw HJ. All-single-mode fiber resonator. Optics Letters. 1982;7:288-290

[13] Stokes LF, Chodorow M, Shaw HJ. All-fiber stimulated Brillouin ring laser with sub-milliwatt pump threshold. Optics Letters. 1982;7:509-511

[14] Jackson DA, Jones JDC. Fiber-optic sensors. Optica Acta. 1986;**33**:1469-1503

[15] Zarinetchi F, Smith SP, Ezekiel S. Stimulated Brillouin fiber-optic laser gyroscope. Optics Letters. 1991;**16**: 229-231

[16] Hill KO, Kawasaki BS, Johnson DC. CW Brillouin laser. Applied Physics Letters. 1976;**28**:608-609

[17] Kalli K, Jackson DA. Ring resonator optical spectrum analyzer with 20 kHz resolution. Optics Letters. 1992;**17**: 1090-1092

[18] Tai S, Kyuma K, Nakayama T. Novel measuring method for spectral linewidth of laser diodes using fiber-optic ring resonators. Electronics Letters. 1985;**21**:91-93

[19] Newton SA, Howland RS, Jackson KP, Shaw HJ. High speed pulse-train generation using single-mode fiber recirculating delay lines. Electronics Letters. 1983;**19**:757-758

[20] Rabus D. Integrated Ring Resonators. New York: Springer; 2007

[21] Heebner JE et al. Optical Micro-Resonators: Theory, Fabrication and Applications. London: Springer; 2008

[22] Heebner JE, Wong V, Schweingsberg A, Boyd RW, Jackson DJ. Optical transmission

characteristics of fiber ring resonators. IEEE Journal of Quantum Electronics. 2004;**40**:736-730

[23] Vien V. Optical Micro-Ring Resonators: Theory, Techniques and Applications. Boca Raton, FL: CRC Press; 2017

[24] Bogaerts W et al. Silicon micro-ring resonators. Laser and Photonics Reviews. 2012;**6**(1):47-53

[25] Knox RM, Thoulios PP. Integrated circuits for the millimeter through optical frequency range. In: Proceedings of Symposium on Submillimeter Waves. Vol. 20; Brooklyn-New York; 1970. pp. 497-515

[26] Marcatili EA. Dielectric rectangular waveguide and directional coupler for integrated optics. The Bell System Technical Journal. 1969;**48**:2071-2102

[27] Mode Solutions-Waveguide Mode Solver and Propagation Simulator. URL: http://www.lumerical.com/mode

[28] WGMODES-Photonics Research Laboratory. URL: http://www.photonics.umd.edu/software/wgmodes/

[29] COMSOL Multiphysics. URL: http://comsol.com/mode

[30] Phoenix Software. URL: http://www.phoenixbv.com/FieldDesigner

[31] Yariv A. Coupled-mode theory for guided-wave optics. IEEE Journal of Quantum Electronics. September 1973; **QE-9**(9):919-9333

[32] RSoft Products-Synopsys Optical Solutions. URL: http://optics.synopsys.com/RSoft/BeamPROP

[33] Phoenix Software. URL: http://www.phoenixbv.com/OptoDesigner

[34] RSoft Products-Synopsys Optical Solutions. URL: http://optics.synopsys.com/RSoft/FemSIM

[35] FDTD Solutions-Lumerical's Nanophotonic FDTD Simulation Software. URL: http://lumerical.com/fdtd

[36] RSoft Products-Synopsys Optical Solutions. URL: http://optics.synopsys.com/RSoft/FullWAVE

[37] Feit MD, Freck JA Jr. Light propagation in graded-index optical fibers. Applied Optics. 1978;**17**:3990-3998

[38] Thylén L. The beam propagation method: An analysis of its applicability. Optics Quantum Electron. 1983;**15**:433-439

[39] Yamauchi J, Shibayama J, Nakano H. Beam propagation method using Padé approximant operators. Transactions IEICE Japan. 1994;**J-77-C-I**:490-494

[40] Hadley GR. Wide-angle beam propagation using Padé approximant operators. Optics Letters. 1992;**17**:1426-1428

[41] Liu PL, Li BJ. Study on form birefringence in waveguide devices using the semi-vectorial beam propagation method. IEEE Photonics Technology Letters. 1991;**3**:913-915

[42] Koshiba M, Tsuji Y. A wide-angle finite element beam propagation method. IEEE Photonics Technology Letters. 1996;**8**:1208-1210

[43] Zienckiewitz OC. The Finite Element Method. 3rd ed. New York: McGraw-Hill; 1973

[44] Koshiba M, Saitoh H, Eguchi M, Hirayama K. Simple scalar finite element approach to optical waveguides. IEEE Proceedings Journal. 1992;**139**:166-171

[45] Koshiba M. Optical Waveguide Theory by the Finite Element Method. Dordecht, Holland: KTK Scientific Publishers and Kluwer Academic Publishers; 1992

[46] Meunier G. The Finite Element Method for Electromagnetic Modeling. River Street-Hoboken NJ: John Wiley & Sons; 2008

[47] Kwon YW, Bang H. The Finite Element Method Using Matlab. CRC Press; 1977

[48] Stern M. Semivectorial polarized finite difference method optical waveguides with arbitrary index profiles. IEEE Proceedings Journal. Piscataway NJ (USA); 1988;**135**:56-63

[49] Stern M. Semivectorial polarized H-field solutions for dielectric waveguides with arbitrary index profiles. IEEE Proceedings Journal. Piscataway NJ (USA); 1988;**135**:333-338

[50] Kawano K, Kitoh T, Kohtuku M, Takeshita T, Hasumi Y. 3-D semivectorial analysis to calculate facet reflectivities of semiconductor optical waveguides based on the bi-directional method of line BPM (MoL-BPM). IEEE Photonics Technology Letters. 1998;**10**: 108-110

[51] Gerdes B, Lunitz B, Benish D, Pregla R. Analysis of slab waveguide discontinuities including radiation and absorption effects. Electronics Letters. 1992;**28**:1013-1015

[52] Taflove A. Computational Electrodynamics: Finite Difference Time Domain Method. Artech House: Norwood, MA; 1995

Chapter 5

Modal Interferometers Based on a Tapered Special Photonic Crystal Fiber for Highly Sensitive Detection

Vladimir (or Uladzimir) P. Minkovich, Joel Villatoro and Pavel V. Minkovich

Abstract

The use of a tapered special photonic crystal fiber (PCF) with collapsed air holes in the waist (the thinnest part of a taper) for highly sensitive detection of strain, high temperature, and fast detection of hydrogen with concentrations between 1.2 and 5.6 vol.% and biosensing is demonstrated. In the tapered PCF, a fundamental core mode couples to a few modes of the solid taper waist. Owing to the beating between the waist modes, the transmission spectra of the tapered PCF exhibit several interference peaks, which are sensitive to refractive index changes of a medium that surrounds the taper and also to changes of a taper length. The changes can be visualized as a shift of the peaks in the output spectrum pattern.

Keywords: modal interferometer, optical fiber devices, optical fiber sensors, interferometry, photonic crystal fiber

1. Introduction

Photonic crystal fibers (PCFs), also known as microstructured or holey optical fibers [1–4], consist of a waveguiding core surrounded by a system of air channels in glass cladding that run along the length of the fiber and arranged in a hexagonal structure around the core. The structure of PCFs enables new possibilities for optical sensing in comparison with standard optical fibers. The most common approach consists of making a sample to interact with the evanescent fields of the PCF guided modes [5–7]. To do so one has to fill the holes with the sample, a gas or liquid, for example, and then the analysis or detection is carried out. In some situation such a process may be inconvenient or impractical. Mach-Zehnder modal interferometers (MZMIs) based on no adiabatic tapered silica PCFs, first fabricated in our facilities [8–10], are attractive for sensing application because of their intrinsic advantages, such as simplicity of fabrication and practical using, high sensitivity, small size, and immunity to electromagnetic interference and their aptitude for remote measurements. Initially, our MZMIs were used for high-resolution refractive index sensing of liquids with indices ranging from 1.41 to 1.45 [8]. In this work, it is presented fabrication such MZMIs and also their application for detection (with a very high sensitivity) of strain, high temperature, and fast detection of hydrogen with concentrations between 1.2 and 5.6 vol.% or for biosensing.

2. PCF taper fabrication and operating mechanism

To fabricate tapers we employed a homemade quasi-single-mode large-mode-area PCF consisting of a solid core surrounded by four rings of air holes in a cladding arranged in a hexagonal pattern [11]. A micrograph of a cleaved end face of our PCF before the tapering is shown in **Figure 1** (left). The parameters of our PCF are outer diameter of 125 μm, core diameter of 11 μm, average hole diameter of 2.7 μm, and average hole spacing (pitch) of 5.45 μm. To reduce the PCF diameter, the fiber is heated with an oscillating high-temperature flame torch and slowly stretched ("slow-and-hot" method). A Vytran GPX3400 glass-processing machine also can be used. At such a no adiabatic tapering process, a waist of the PCF (the thinnest part of a taper) can be reduced until the air holes collapse, obtaining a piece of a solid fiber with the diameter $\rho_W \sim <33$ μm, which can support multiple modes (waist modes). In the first transition zone, see **Figure 1** (right), some energy from the PCF fundamental mode passes to a few waist modes.

The beating of these modes inside the solid waist section is sensitive to an external environment, since the propagation constants of the waist modes depend on it. In the final transition zone, the waist modes are again transformed into the PCF fundamental mode, which intensity is determined by the phase difference between the waist modes [9]. The resulting mode carries the interference information generated at the tapered zone to a detector. If, for the simplicity, we consider only two waist modes for the interference with effective refractive indices n_1 and n_2, then, intensity output is

$$I_t = I_1 + I_2 + 2\sqrt{I_1 I_2}\cos(\Delta\theta), \quad (1)$$

where I_1 and I_2 are the intensities of the interactive waist modes, respectively, and $\Delta\theta$ is the phase difference between them. This phase difference creates the output pattern and depends on the difference between the effective waist mode refractive indices (Δn) and the length of the waist L_0:

$$\Delta\theta = \frac{2\pi}{\lambda}(\Delta n)\, L_0, \quad (2)$$

where λ is the center wavelength of the light source used. The interferometer transmittance maxima will be at $2\pi\Delta n L_0/\lambda = 2\pi m$, where m is an integer. Peaks in the output spectrum pattern will be appeared at wavelengths given by $\lambda_m \approx \Delta n L_0/m$, and the spacing between consecutive peaks (period) is provided by $P \approx \lambda^2/\Delta n L_0$. When the refractive index of the surrounding medium is changed, Δn is changed too (keeping L_0 fixed), and a change of phase is generated. This change can be visualized as a peak shift in the output spectrum pattern. The shift in the output spectrum pattern is also happened, when L_0 is changed [9]. It should be pointed out that additional losses, because of PCF tapering, were found to be typically below 3 dB. To test our sensors, we implemented a simple light transmission measuring setup consisting of a low-power light-emitting diode (LED) with peak emission at 1280 nm (or 1550 nm) and about 40–80 nm of spectral width and an optical spectrum analyzer (OSA), Ando AQ-6315E. The tapered PCFs (with a PCF full length of approximately 30 cm) were fusion spliced between standard fibers (SMF-28), and the tapered section was held straight and secured in a mount in all the experiments.

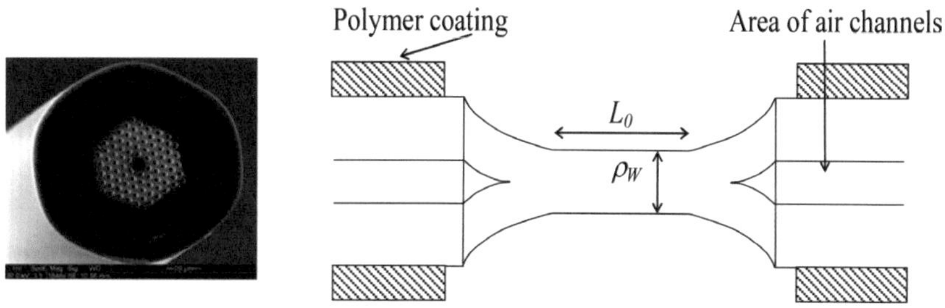

Figure 1.
Image of the cross section of an untapered PCF (left) used to fabricate our tapers and illustration of a uniform-waist tapered PCF (right). L_0 *is the length of the uniform waist, and* ρ_W *is the taper waist diameter.*

3. Strain detection

There are a number of applications in which the monitoring of strain-induced changes is important. Some designs based on standard optical fibers have been reported in the literature. For example, strain sensors based on in-fiber gratings [12] and Brillouin scattering [13] are commercially available. The main shortage of these strain sensors is their high thermal sensitivity [14]. In this case, one needs to simultaneously and independently measure strain and temperature [12, 13]. So, the complexity of the sensor is increased. It is necessary to mention that a few attempts to sense strain with PCFs have been reported in the literature at the beginning of our study. They include the use of Brillouin frequency shift [15], long-period grating [16], and fiber Bragg grating [17, 18]. However, these sensors also exhibit an undesirable cross sensitivity to temperature. The application of our tapered PCFs with collapsed air holes for temperature-independent strain sensing is described in this section [19, 20]. Using the fiber, and the tapering process as described in a previous section, a silica PCF taper with waist diameter ρ_W = 28 μm and L_0 = 5 mm was fabricated. **Figure 2** shows the normalized transmission spectra of the used PCF before (dotted line) and after (continuous line) a no adiabatic tapering process.

The measurements were carried out in a measuring setup consisting of an LED, with peak emission at 1540 and 40 nm of spectral width, and an optical spectrum

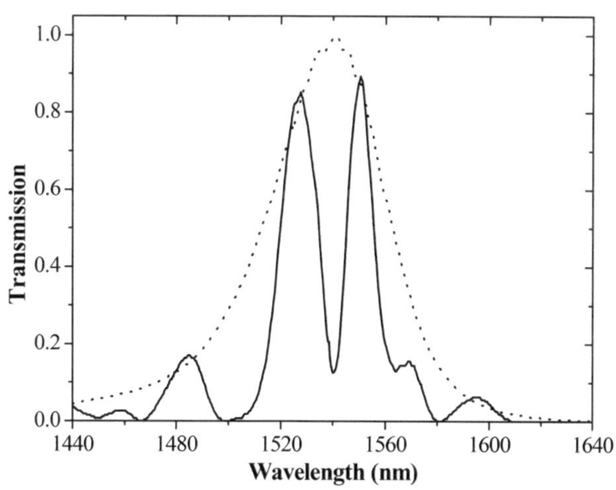

Figure 2.
Normalized transmission spectra of the PCF before (dotted curve) and after (continuous curve) tapering. The taper waist diameter is 28 μm (reprinted with permission from Ref. [19], OSA).

analyzer with a resolution of 0.1 nm. It is possible to note that the transmission of our untapered PCF is basically the output spectrum of the LED used. However, the spectrum of the 28-μm-thick taper exhibits a series of peaks, two of which are higher than the others. For this taper we investigated the shift of the interference peaks caused by longitudinal strain. The PCF was fixed between two displacement mechanical mounts, with the tapered section in the middle. Then the fiber was stretched using the calibrated micrometer screws of the mounts. **Figure 3(a)** shows the normalized spectra, measured at 1540 nm, of the taper, when it has been subjected to 0 (continuous curve), 1100 (dashed curve), and 2200 (dotted curve) με. In this figure we can see shift of the initial spectrum to shorter wavelengths, when the strain is increased. When the strain was removed from the sensor, all the peaks returned to their baselines. At this point, we exchanged the LED for another with peak emission at 1300 nm and repeated the experiments. **Figure 3(b)** shows the normalized spectra, measured at 1300 nm, of the taper, when it also subjected to three applied strains: 0 (continuous curve), 1100 (dashed curve), and 2200 (dotted curve) με. It is possible to see in this figure that the transmission spectrum of the device also exhibits interference peaks near 1300 nm and that such peaks also shift to shorter wavelengths. Note from **Figure 3** that the height of some peaks increases and that of others decreases. All the peaks, however, maintain almost the same shape. The influence of temperature on the peaks was also investigated. The taper subjected to 0 με was exposed to different temperatures between 0 and 180°C. In that range of temperatures the interference peaks did not suffer any shift, but, at higher temperatures, the peaks shifted to longer wavelength. We did not carry out measurements below 0°C because of technical limitations. Hence, a no adiabatic tapered silica PCF with collapsed air holes can be used for temperature-independent strain sensing.

The advantage of the sensor is that one can monitor one or all the peaks. In addition, different wavelengths can be used to interrogate the sensor. **Figure 4** shows the peak shift as a function of the applied strain, when the initial peaks are centered around 1520 and 1250 nm, see **Figure 3(a)** and **(b)**, respectively. It is possible to

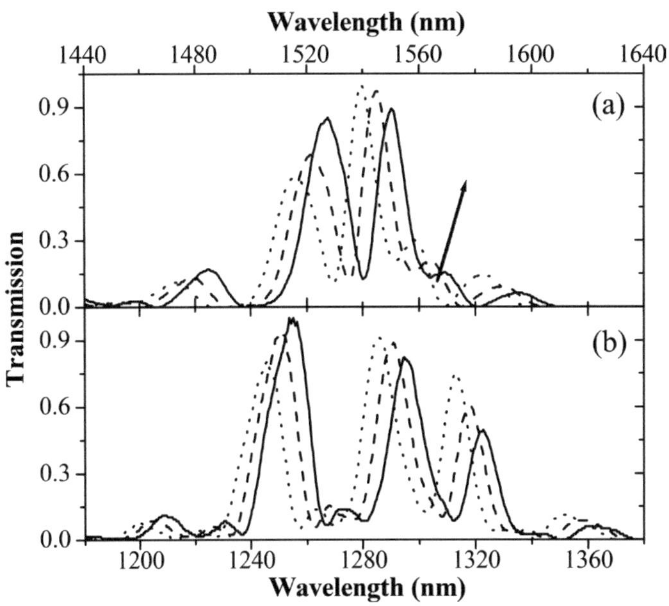

Figure 3.
Normalized transmission spectra of a 28-μm-thick tapered PCF under three applied strains measured at (a) 1550 and (b) 1330 nm. In both figures the continuous curves represent 0 με, the dashed curves 1100 με, and dotted curves 2200 με (reprinted with permission from Ref. [19], OSA).

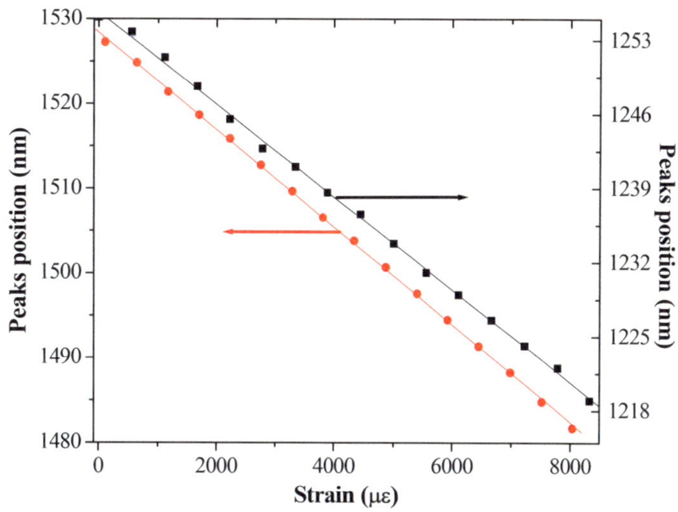

Figure 4.
Typical shift of the peaks as a function of the applied strain (reprinted with permission from Ref. [19], OSA).

note from the figure that the shift of both peaks has a linear behavior and the slope of both lines is basically the same. The observed shift of the other peaks shown in **Figure 3** was also linear, and their slope was similar to that of the peaks shown in **Figure 4**. The experiments were carried out several times. It was observed that in all cases the sensor was reversible in the 0–8000 με range.

It is important to point out that optical fiber interferometric strain sensors are useful devices, because they can provide important information or solutions in a number of applications of practical interest. These sensors, incorporated into civil aircraft and spacecraft structures, smart materials, active devices and components, etc., permit the monitoring of strain-induced changes suffered by such structures, materials, or components. In all these applications, temperature-independent, intrinsic, and wavelength-encoded strain sensors with high resolution are desirable.

4. High-temperature detection

It is known variety of standard fiber-based temperature sensors, for both point and distributed detection. The sensing mechanisms include [21] fluorescence and time decay effects in active materials and doped fibers, remote blackbody radiation, Raman and Brillouin scattering, interferometry, and Bragg or long-period grating technology. The large majority of reported so far fiber temperature sensors have been designed to operate in a range from −20 to 200°C. However, there are some applications in which high-temperature sensing is necessary, for example, for monitoring furnace operation or volcanic events, or in fire alarm systems, etc. [22]. Most of the techniques mentioned in this chapter before are not suitable for high-temperature sensing. Materials traditionally used for fluorescence-based fiber thermometers have an inferior fluorescence intensity emission at temperatures above 600°C [23, 24]. Some interferometric temperature sensors can be designed for measurement of temperatures higher than 1600°C [25, 26]. However, their construction is complicated and a sensing element is external to the fiber. Important advances have been made to fabricate fiber Bragg grating devices for measurement of high temperatures [23], but these devices require a complicated fabrication process or a long and controlled temperature treatment [27–29]. A long-period grating inscribed with a pulsed CO_2 laser in a PCF [30] or a very small stub of index-guiding PCF fusion spliced between two standard single-mode fibers [31] can be also used for high-temperature measurements. We have proposed a novel high-temperature sensor based on the developed PCF taper with collapsed air holes in the waist [32, 33]. The PCF taper with ρ_w = 31 μm and L_0 = 5 mm (see **Figure 1**) was used for high-temperature measurements. Before the tapering, a few centimeters of the referred PCF were inserted between two standard single-mode fibers (SMF-28) by fusion splicing. Then the PCF was slowly stretched, while a section of length L_0 was heated at a high temperature (at about 1000°C). It is important to point out that the PCF can also be tapered without the need of splicing it with conventional optical fibers. The fabricated taper was placed within a pure silica capillary tube, in order to have the taper straight during experiments (bending affects the sensor response). Then, the whole set was placed into a temperature chamber. To interrogate the device, light was injected from an LED with a peak emission at 1290

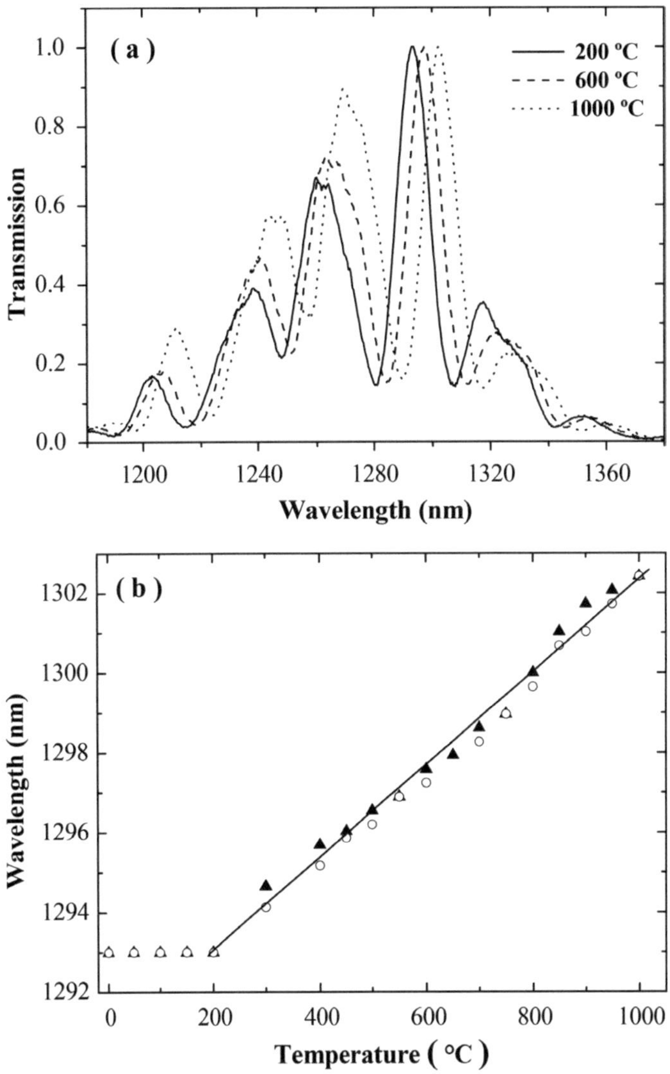

Figure 5.
(a) Normalized transmission spectrum of a taper with ρ_w = *31 μm and* L_0 = *5 mm, at different temperatures. (b) Position of the highest peak of the spectrum (a) versus temperature. The filled triangles (heating) and the hollow circles (cooling) are experimental values. The solid line is a linear fit to the data in the 200–1000°C range (reprinted with permission from Ref. [32], IEEE).*

and with 80 nm of spectral width. The output spectrum was monitored with an optical spectrum analyzer (OSA) with a resolution of 0.1 nm. We also realized our experiments using an LED, with a center wavelength at 1520 nm and with 40 nm of spectral width. The heating or cooling of the taper in the furnace was conducted in steps of 50°C. In each step of heating or cooling, one waited for about 30 min before collecting any data. With this procedure a homogeneous temperature inside the chamber was ensured. The exhibited around 1300 nm transmission spectra of the referred taper at temperatures of 200°C (solid line), 600°C (dashed line), and 1000°C (dotted line) are shown in **Figure 5(a)**. Measurements for temperatures above 1000°C were not possible for limitations of our furnace. In the plots of **Figure 5(a)**, we can see that the interference peaks shift toward longer wavelengths as the temperature increases. We have also found that the shift of the peaks shown in **Figure 5(a)** using the LED centered at 1290 nm and the shift of the peaks using the LED centered at 1520 nm (is not shown) are similar. This means that the central peak of the LED is not important. One also can see that the peaks are not deformed, when the device is exposed to a temperature higher than 1000°C. This is so, because the fiber was made of pure silica and the taper was fabricated, when the PCF was exposed to a high-temperature flame (of approximately 1000°C). The experiments were repeated several times in a period of 4 weeks, and it was observed the same behavior of the sensor. The position of the highest peak maximum for the spectrum of **Figure 5(a)** versus temperature is shown in **Figure 5(b)**.

One can see that the peaks are insensitive to temperature in the 0–180°C range; but they shift linearly to longer wavelengths in the 200–1000°C range. The slope of the curve in such a range is 12 pm/°C. Deviations from this line were, in our opinion, due to errors in the readings of the furnace temperature and also in the definition of the peak maxima position. The resolution of the sensor can be improved using thinner tapers, which exhibit sharper peaks that are easier to monitor. Unfortunately, thinner tapers are more difficult to work with. Another possibility to improve the sensor resolution is to monitor all the interference peaks and to combine them with a fitting algorithm.

5. Gas detection

To the beginning of our study, several sensors that exploit the direct interaction of evanescent fields of PCF guided modes with the target gas within the holes of unmodified PCFs have been proposed and demonstrated [5, 34–37]. The walls of the PCF air holes can also be covered with thin layers for a selective detection of specific gases [38]. Unfortunately, the filling of the air holes with a target gas is not convenient enough, since it takes a long time. Theoretical and experimental studies have shown that the time for a gas to diffuse into the very small holes of a PCF takes about tens of seconds to several minutes and depends on the fiber length [34]. To the time needed for a gas to diffuse into the holes of a PCF, one has to add the time needed to detect, analyze, and process the signal. Leaving the microscopic holes open in a PCF is also not convenient, since they can be filled with undesirable microparticles or moisture that can block the holes or change an output signal. In some potentially explosive or flammable environment, for example, in hydrogen environments, fast gas detection is necessary, and a rapid response of the sensor is desirable [39]. It is necessary also to remind that even existent this time hydrogen sensors have the response time not fast enough [40, 41]. We proposed the use of developed tapered PCFs, coated with thin layers, which can absorb the sensing gas or chemicals, for faster detection. In the taper waist (see **Figure 1**), the external medium plays the role of cladding, and the solid waist section plays the role of core. Therefore, a thin layer deposited on a waist region will attenuate or absorb the evanescent fields of different propagating waist modes. As a result, the output pattern of the tapered fiber will be also modified. Thus, the sensing of different gases or any other chemicals is observable. To confirm of the principle, we demonstrated the sensing of hydrogen [42, 43]. Hydrogen is one of the cleanest energy sources. It can be used in many chemical processes and in various fields, for example, as propellant in aerospace rockets, fuel for fuel cells, or engines in automotive devices. However, hydrogen is extremely flammable and can be explosive in air at room temperature and pressure even at concentrations of 4 vol.% and of course at more ones [39]. Therefore, fast detection of hydrogen at low concentrations is necessary. To do so, an ultrathin palladium layer was deposited on a solid section of a tapered PCF. It is well known that a thin palladium (or palladium alloy or composite) film can selectively and reversibly absorb hydrogen [39, 44–51]. When a Pd or Pd-alloy film is exposed to hydrogen, it is converted to PdH. The hydration of the thin film makes the real and imaginary part of the film dielectric constants to decrease [45, 49]. Such changes in the Pd film are possible to monitor with optical methods [39, 44–51].

Using the same fiber, and the same tapering process as in previous sections, a silica PCF taper with waist diameter ρ_w = 28 μm and L_0 = 10 mm was fabricated. An 8-nm-thin film was deposited on the waist of the tapered PCF over a length of 10 mm. The palladium film was deposited in a high-vacuum chamber by thermal evaporation. The sensor was tested (at normal conditions) in a gas chamber, which had an inlet and outlet to allow the hydrogen or the hydrogen/carrier mixture to flow in and out. Nitrogen was used as a carrier gas. The flow of nitrogen and hydrogen was controlled individually with mass flow controllers. The sensor is tested in a transmission measurement setup consisting of a low-power LED with peak emission at 1290 and 80 nm of spectral width and an OSA. The spectra of the device were recorded and analyzed at different concentrations of hydrogen. Before measurements, about 30 cm of the PCF was fusion spliced between two standard single-mode fibers to decrease the sensor cost, and then the tapered section was held straight in a mechanical mount during all the experiments. Results of our experiments are presented in **Figure 6**. The top side plots in **Figure 6** show the spectra that were received, when the sensor was exposed to hydrogen concentrations between 1.2 and 5.6%. Four peaks of the spectra are numbered for convenience. Intensity of the peaks 1–4 as a function of hydrogen concentrations is presented in the bottom side of **Figure 6**. One can see that the intensity of the peaks increases in a nonlinear manner with the increase of the hydrogen concentration. For hydrogen concentrations, more than 6% the sensor exhibited spectra similar to that at 5.6%, which indicated the saturation of the 8-nm-thick palladium film. The reason of increase of the obtained peak intensity is because the index of the thin palladium film decreases, when it is exposed to hydrogen [45, 47, 49]. Such a decrement of the index causes the absorption of the evanescent fields to decrease. The remarkable increment of the peak intensity demonstrates the high sensitivity of the sensor. The sensor response time (the time required for the sensor to reach 90% of transmission change) was approximately 10 s. In **Figure 6** one can see that the intensity changes are more in the peaks or maxima than in the valleys or minima. One can also note that the intensity changes at longer wavelengths, see peaks 3 and 4, are more pronounced than at shorter wavelengths, see peaks 1 and 2. This is due to the fact that the attenuation of the palladium film increases, when the wavelength augments, according to the increase of the evanescent fields [47].

The interferometric hydrogen sensor reported here is more compact, simpler, and also faster than other interferometer-based hydrogen sensors reported in the literature [39, 44–46, 50].

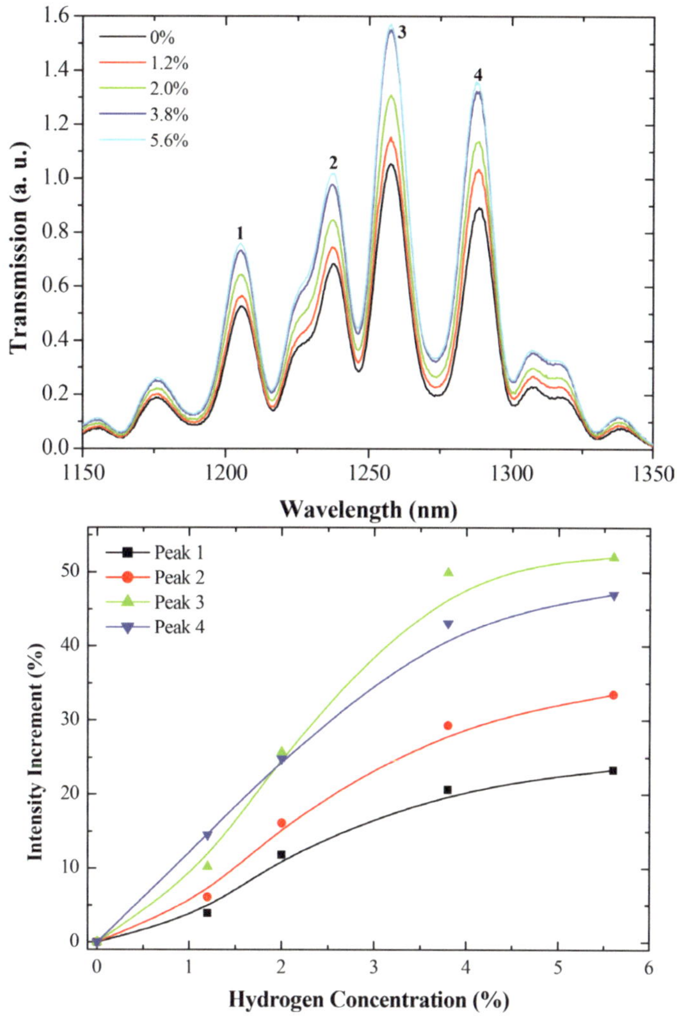

Figure 6.
Transmission spectra of a tapered PCF coated with an 8-nm-thick Pd film at different hydrogen concentrations (top) and relative increment of the peaks 1–4 as a function of hydrogen concentration (bottom). Parameters of the sensor: ρ_w = *28 μm,* L_o = *10 mm (reprinted with permission from Ref. [42], OSA).*

6. Highly sensitive biosensing

Previous experimental studies of modal interferometers, for the biosensing refractive index (RI) range (1.33–1.34), have informed about sensitivities of 320 nm/RIU for a sensing length of PCF (~3.0 mm) [52] and sensitivities of 1629 nm/RIU for a longer sensing length (~24 mm) [53]. PCF MZMIs based on no adiabatic tapered fibers are attractive for biosensing applications, because they can have a very small sensing length [8, 10]. Unfortunately, PCF MZMI RI sensors for biosensing applications usually can only be used to detect a solution with one analyte and when the RI dependence of an analyte at different concentrations is known. To obtain a specific sensitivity for a chosen analyte in some complex solutions, we coated the sensing surface (tapered PCF) of our RI sensor with a layer of an active biological component (an immobilization method), characterized by having a high affinity with the chosen analyte. The usual immobilization method, for glass substrates, is through the process of silanization, followed by the covalent bonding of an antigen or antibody [52, 54]. In [52], a detection limit of 10 μg/ml was demonstrated for streptavidin using a modal interferometer based on a short piece of PCF (~3.0 mm) spliced between two standard single-mode fibers, while in [55], a modal interferometer based on a thin-core fiber spliced between two standard single-mode fibers was used, and a detection limit of 1.1 ng/ml was achieved. In this paragraph, we inform about a MZMI based on a no adiabatic tapered special silica PCF acting as a highly sensitive biosensor. We also demonstrated that as a refractometer, it has good sensitivity in the RI range of interest for biological solutions and a record detection limit of 125 pg/ml of a protein concentration as a biosensor [56, 57]. In our experiments we used a homemade large-mode-area photonic crystal fiber. A cross section of the untapered fiber is shown in **Figure 7(a)**. The used fiber has an average hole spacing (pitch) Λ of 5.45 μm, an average hole diameter d of 2.7 μm, and an external diameter of 125 μm. The modal properties of the fiber are explained in [11]. To reduce our sensor cost, we spliced about 30 cm of the PCF between two standard SMF-28 single-mode fibers. The difficulty in splicing the PCF with the used standard SMF is that each one requires different temperatures to reach the needed fusion conditions. A commercial splicing machine, Sumitomo Type-46S, was used in the process. The splicing configuration is shown in **Figure 7(b)**. We used a Vytran GPX3400 glass-processing machine to make the no adiabatic tapering. The tapered fiber for our sensor was fabricated with both transition regions (down taper and up taper) being symmetrical with a length of 5 mm, a waist length of 10 mm, and a waist diameter of 18.1 μm, like the one presented as an example in [58] on a base of a standard fiber. The pulling speed of the fiber holding blocks was kept at a rate of 1 mm/s, while the heat was set at 90 W. We used an LED source with a center wavelength of 1550 nm and an optical spectrum analyzer (OSA), Ando AQ-6315E, which has a resolution of up to 0.05 nm, to make the SM-PC-SM fiber transmission spectrum measurements, as shown in **Figure 8(a)**. To make it easier to measure aqueous solutions with the fiber sensor, a work cell, shown in **Figure 8(b)**, was designed.

The cell was fabricated with a working volume of 50 μl and a cavity length of 13 mm. To test a sensitivity of our device for biosensing RI range, solutions of sodium chloride (NaCl) diluted in distilled water were prepared in the following concentrations, 0.0, 0.2, 0.4, 0.6, 0.8, and 1 M, with corresponding refractive indices at 1550 nm of 1.30864, 1.31104, 1.31339, 1.31569, 1.31794, and 1.32014 [59]. An antigen-antibody test was proposed to evaluate the performance of the device for a biosensor application. We chose the bovine serum albumin (BSA) antigen because of its low cost, ease of handling, availability, and immobilization (covalent binding between the molecule and the transducer surface), which has been widely studied. The immobilization process was conducted like the

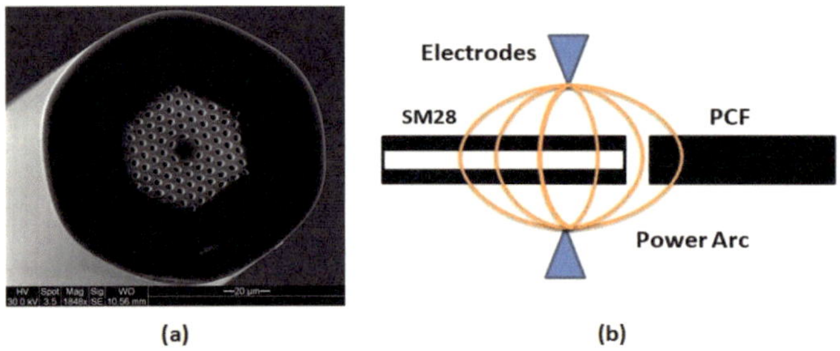

Figure 7.
(a) PCF untapered cross section, (b) SM-PCF splicing configuration (reprinted with permission from Ref. [56], IEEE/OSA).

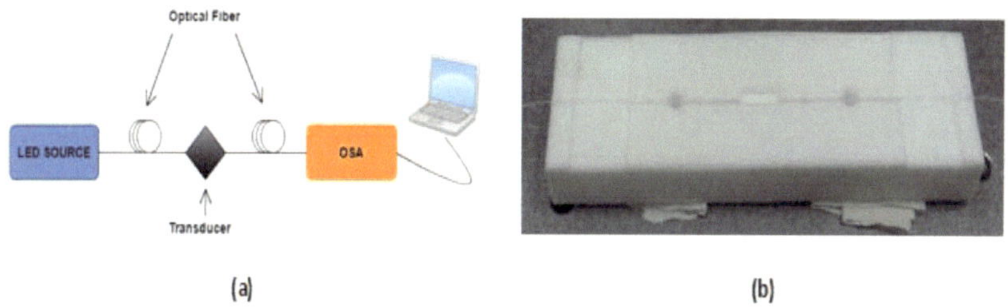

Figure 8.
(a) Experimental setup, (b) work cell (reprinted with permission from Ref. [56], IEEE/OSA).

one presented in [54]. Preliminary, the activation of the taper surface with an aminosilane APTES was made, and then the BSA antigen was covalent coupled to the taper surface. After the antigen-immobilization process, the performance tests were conducted by passing the respective antibody (anti-BSA) diluted in phosphate-buffered saline (PBS) buffer at different concentrations (125, 12.5, 1.25, and 0.125 ng/ml) over the sensitive sensor's surface. All tests were carried out at room temperature, at about 25°C. The protocol used for the tests was as follows. Step 1: Apply tris(hydroxymethyl)aminomethane (TRIS) buffer for 3 min to clean and break previous bonds. Step 2: Remove the TRIS buffer, apply the PBS, and make a measurement of the optical spectrum that corresponds to the baseline. Step 3: Remove the PBS buffer, apply the desired antibody-PBS sample, and take measures for 10 min. Previous experiments have demonstrated that after 6 min of interaction, the variation of the signal can be neglected for practical measurements. For each sample, repeat the protocol. The difference in wavelength between a peak on the baseline and the same peak, measured after 8 min of sample interaction, is the biosensor response, which is directly related to the sample concentration. **Figure 9(a)** shows that as the refractive indices of the NaCl solutions increase, the peaks shift to longer wavelengths. Peaks with similar morphologies were selected for the analysis of the peak shifts. **Figure 9(b)** corresponds to the linear fit built by taking the wavelength of the peak maxima (between 1538 and 1548 nm) and correlating them with the respective RI. A sensitivity of 722.3 nm/RIU was achieved (with a linearity of 0.9965), which can be compared with other sensors reported. **Figure 10** shows the results of the biosensor application, where the magnitude of the response is directly proportional to the anti-BSA concentration.

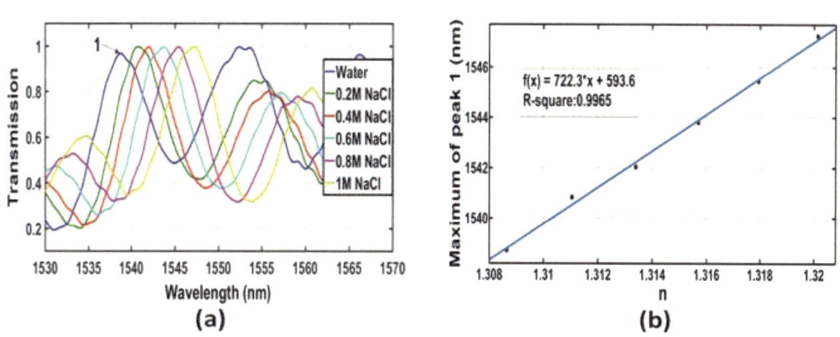

Figure 9.
(a) Transmission spectra of the sensor at different concentrations of NaCl, (b) linear fit of the sensor response for concentrations of NaCl with RI between 1.30864 and 1.32014 (reprinted with permission from Ref. [56], IEEE/OSA).

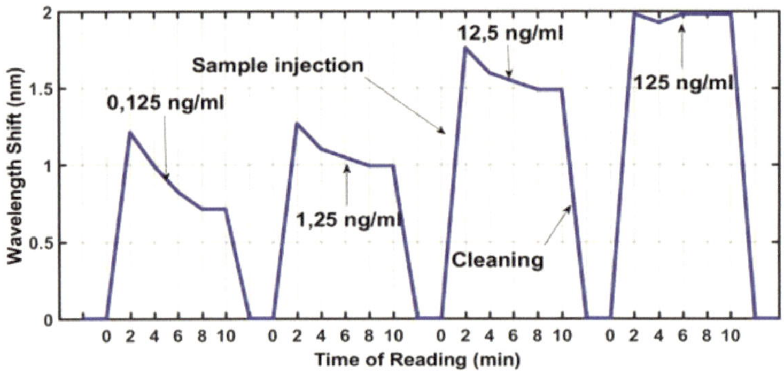

Figure 10.
Sensor response for concentrations of anti-BSA between 0.125 and 125 ng/ml (reprinted with permission from Ref. [56], IEEE/OSA).

We can see in **Figure 10** that the sensor can take measurements of concentrations as low as 125 pg/ml; this means that our detection limit is lower than the one reported in [58]. Another fact to highlight is that our sensor is capable to detect a specific protein inside a complex sample (with different proteins diluted) that is not possible for the sensor proposed in [58]. The increased sensitivity in our case is possible, when compared with that in [58], because in [58] the detection of protein concentrations was done by reading a change in the refractive index of the bulk solution; therefore, higher concentrations were needed to produce a change big enough to be detected by the sensor. In our case, the antigen of the target molecule was immobilized onto the transducer surface, making it possible to detect the antigen-antibody interaction with higher sensitivity. The changes in refractive index near the surface are mostly affected by the coupling of the specific protein and not by other proteins in the sample. In addition, the area near the surface of the fiber has exponentially higher sensitivity than areas that are farther away. The developed sensor also has a lower detection limit when compared with those reported in [52, 55], with detection limits of 10 and 1.1 ng/ml, respectively, outperforming the former in three orders of magnitude and the latter in one order of magnitude. It is also necessary to point out that the estimated maximum resolution of our sensor was found to be around 1×10^{-2} ng/ml, considering that the resolution of the spectrum analyzer used was 0.05 nm. The higher resolution of our measurements is possible at using an OSA with higher resolution. More precise localization of interference fringes by performing a Fourier transformation will be also helpful [60]. As can be also seen in **Figure 10**, we did not have any saturation in a sensor response for concentrations of anti-BSA between 0.125 and 125 ng/ml, although the spectral shift had highly nonlinear dependence on the anti-BSA concentration. Comparing with **Figure 10**, where we had a linear dependence of our sensor response and did not have any chemical binding between NaCl solutions and the taper surface, one can conclude that the antibody-antigen binding process has a great impact on our sensor response. A definition of the impact requires further investigations.

7. Conclusions

In this chapter, a special homemade quasi-single-mode PCF was used for fabrication of no adiabatic tapered fibers with a solid waist. The fabrication of tapers is simple and takes only few minutes. It is also possible to control the process of tapering at all times. It has been found that in PCF tapers with waist diameters small enough (thinner than 33 μm), the air holes of a PCF cladding were collapsed. A waist of such tapers consists of a solid multimode fiber supporting multiple modes. The interference between such modes gives series of peaks in an output spectrum pattern. The peaks are sensitive to a medium that surrounds the taper and also to the length of the waist, since the propagation constants of the interfering modes depend on them. Any changes in a surrounding medium or in the length of the taper are visualized as a shift of peaks in the output spectrum pattern. The tapers were used for fabrication of very sensitive interferometric devices with selective transmission for temperature-insensitive detection (in the 0–180°C range) of microstrain, for sensing of high temperature in the 200–1000°C range, and for fast detection of hydrogen concentrations between 1.2 and 5.6 vol.%. Moreover, the developed interferometers can be used for detection of interaction between a BSA antigen and an anti-BSA antibody with a record detection limit of 125 pg/ml of antibody concentration. The parameters that are necessary to measure by using the interferometers are transformed into the shift of peaks in the output spectrum pattern. It is not difficult by using an optical spectrum analyzer to monitor the shift. Detection is fast and takes approximately 10 s. It is necessary also to note that the output spectrum patterns are stable and immune to possible fluctuations of a light source power.

Acknowledgements

The authors of the chapter would like to thank their colleagues and co-authors of joint publications Monzon-Hernandez D, Kir'yanov AV, Calixto S, Sotsky AB, Sotskaya LI, Badenes G, Betancur-Ochoa JE, and Montagut-Ferizzola YJ for long-term and productive scientific cooperation.

Author details

Vladimir (or Uladzimir) P. Minkovich[1*], Joel Villatoro[2,3] and Pavel V. Minkovich[4]

1 Division of Photonics, Centro de Investigaciones en Optica, Leon, GTO, Mexico

2 Department of Communications Engineering, University of the Basque Country, Bilbao, Spain

3 Ikerbasque-Basque Foundation for Science, Bilbao, Spain

4 Mogilev State Medical College, Mogilev, Belarus

*Address all correspondence to: vladimir@cio.mx

References

[1] Knight JC, Birks TA, Russell PSJ, Atkin DM. All-silica single-mode optical fiber with photonic crystal cladding. Optics Letters. 1996;**21**:1547-1549. DOI: 10.1364/OL.21.001547

[2] Russell P. Photonic crystal fibers. Science. 2003;**299**:358-362. DOI: 10.1126/science.1079280

[3] Bjarklev A, Broeng J, Bjarklev AS. Photonic Crystal Fibres. Boston, MA: Kluver Academic Publishers; 2003. 298 p. ISBN: 1-4020-7610-X

[4] Russell PSJ. Photonic crystal fibers. Journal of Lightwave Technology. 2006;**24**:4729-4749. DOI: 10.1109/JLT.2006.885258

[5] Monro TM, Belardi W, Furusawa K, Baggett JC, Broderick NGR, Richardson DJ. Sensing with microstructured optical fibers. Measurement Science and Technology. 2001;**12**:854-858. DOI: 10.1088/0957-0233/12/7/318

[6] Konorov SO, Zheltikov A, Scalora M. Photonic-crystal fiber as a multifunctional optical sensor and sample collector. Optics Express. 2005;**13**:3454-3459. DOI: 10.1364/OPEX.13.003454

[7] Cubillas AM, Unterkofler S, Euser TG, Etzold BJM, Jones AC, Sadler PJ, et al. Photonic crystal fibres for chemical sensing and photochemistry. Chemical Society Reviews. 2013;**42**:8629-8648. DOI: 10.1039/C3CS60128E

[8] Minkovich VP, Villatoro J, Monzón-Hernández D, Calixto S, Sotsky AB, Sotskaya LI. Holey fiber tapers with resonance transmission for high-resolution refractive index sensing. Optics Express. 2005;**13**:7609-7614. DOI: 10.1364/OPEX.13.007609

[9] Minkovich VP, Monzón-Hernández D, Villatoro J, Sotsky AB, Sotskaya LI. Modeling of holey fiber tapers with selective transmission for sensor application. Journal of Lightwave Technology. 2006;**24**:4319-4326. DOI: 10.1109/JLT.2006.884207

[10] Villatoro J, Minkovich VP, Monzon-Hernandez D. Compact modal interferometer built with tapered microstructured optical fiber. IEEE Photonics Technology Letters. 2006;**18**:1258-1260. DOI: 10.1109/LPT.2006.875520

[11] Minkovich VP, Kir'yanov AV, Sotsky AB, Sotskaya LI. Large-mode-area holey fibers with a few air channels in cladding: Modeling and experimental investigation of the modal properties. Journal of the Optical Society of America B: Optical Physics. 2004;**21**:1161-1169. DOI: 10.1364/JOSAB.21.001161

[12] Kersey AD, Davis MA, Patrick HJ, LeBlanc M, Koo KP, Askins CG, et al. Fiber grating sensors. Journal of Lightwave Technology. 1997;**15**:1442-1463. DOI: 10.1109/50.618377

[13] Ohno H, Naruse H, Kihara M, Shimada A. Industrial applications of the BOTDR optical strain sensors. Optical Fiber Technology. 2001;**7**:45-64. DOI: 10.1006/ofte.2000.0344

[14] Villatoro J, Finazzi V, Minkovich VP, Pruneri V, Badenes G. Temperature-insensitive photonic crystal fiber interferometer for absolute strain sensing. Applied Physics Letters. 2007;**91**:AD091109. DOI: 10.1063/1.2775326

[15] Zou L, Bao X, Afshar S, Chen L. Dependence of the Brillouin frequency shift on strain and temperature in a photonic crystal fiber. Optics Letters. 2004;**29**:1485-1487. DOI: 10.1364/OL.29.001485

[16] Zhu Y, Shum P, Bay HW, Yan M, Yu X, Hu J, et al. Strain-insensitive and high-temperature long-period gratings inscribed in photonic crystal fiber. Optics Letters. 2005;**30**:367-369. DOI: 10.1364/OL.30.000367

[17] Martelli C, Canning J, Groothoff N, Lyytikainen K. Strain and temperature characterization of photonic crystal fiber Bragg grating. Optics Letters. 2005;**30**:1785-1787. DOI: 10.1364/OL.30.001785

[18] Frazão O, Carvalho JP, Ferreira LA, Araujo FM, Santos JL. Discrimination of strain and temperature using Bragg gratings in microstructured and standard optical fibers. Measurement Science and Technology. 2005;**16**:2109-2113. DOI: 10.1088/0957-0233/16/10/028

[19] Villatoro J, Minkovich VP, Monzon-Hernandez D. Temperature-independent strain sensor made of tapered holey optical fiber. Optics Letters. 2006;**31**:305-307. DOI: 10.1364/OL.31.000305

[20] Minkovich VP, Villatoro J, Monzon-Hernandez D. Method for Monitoring Strain using Tapered Microstructured Optical Fiber. European Patent No. EP1962120B1; 2013

[21] Lopez-Higuera JM, editor. Handbook of Optical Fibre Sensing Technology. Chichester: Wiley; 2002. 795 p. ISBN: 0-471-82053-9

[22] Brambilla G, Kee HH, Prunery V, Newson TP. Optical fibre sensors for earth science: From the basic concepts to optimizing glass composition for high temperature applications. Optics and Lasers in Engineering. 2002;**37**:215-232. DOI: S0143-8166(01)00096-3

[23] Shen YH, Pal S, Mandal J, Sun T, Grattan KTV, Wade SA, et al. Investigation of the photosensitivity, temperature sustainability and fluorescence characteristics of several Er-doped photosensitive fibers. Optics Communication. 2004;**237**:301-308. DOI: 10.1016/j.opticom.2004.04.004

[24] Shen Y, Zhao W, Sun T, Grattan KTV. Characterization of an optical fiber thermometer using Tm^{3+}:YAG crystal, based on the fluorescent lifetime approach. Sensors and Actuators, A: Physical. 2003;**109**:53-59. DOI: 10.1016/j.sna.2003.09.006

[25] Zhu Y, Huang Z, Shen F, Wang A. Sapphire-fiber-based white-light interferometric sensor for high-temperature measurements. Optics Letters. 2005;**30**:711-713. DOI: 10.1364/OL.30.000711

[26] Xiao H, Deng J, Pickrell G, May RG, Wang A. Single-crystal sapphire fiber-based strain sensor for high-temperature applications. Journal of Lightwave Technology. 2003;**21**:2276-2283. DOI: 10.1109/JLT.2003.816882

[27] Lowder TL, Smith KH, Ipson BI, Hawkins AR, Selfridge RH, Schultz SM. High-temperature sensing using relief fiber Bragg gratings. IEEE Photonics Technology Letters. 2005;**17**:1926-1968. DOI: 10.1109/LPT.2005.852646

[28] Forkine M. Thermal stability of chemical composition gratings in fluorine-germanium-doped silica fibers. Optics Letters. 2002;**27**:1016-1018. DOI: 10.1364/OL.27.001016

[29] Trpkovsky S, Kitcher DJ, Baxter GW, Collins SF, Wade SA. High-temperature-resistant chemical composition Bragg gratings in Er^{3+}-doped optical fiber. Optics Letters. 2005;**30**:607-609. DOI: 10.1364/OL.30.000607

[30] Zhu Y, Shum P, Bay HW, Yan M, Yu X, Hu J, et al. Strain-insensitive and high-temperature long-period grating inscribed in photonic crystal fiber. Optics Letters. 2005;**30**:367-369. DOI: 10.1364/OL.30.000367

[31] Coviello G, Finazzi V, Villatoro J, Pruneri V. Thermally stabilized PCF-based sensor for temperature measurements up to 1000°C. Optics Express. 2009;**17**:21531-21559. DOI: 10.1364/OE.17.021551

[32] Monzon-Hernandez D, Minkovich VP, Villatoro J. High-temperature sensing with tapers made of microstructured optical fibers. IEEE Photonics Technology Letters. 2006;**18**:511-513. DOI: 10.1109/LPT.2005.863173

[33] Minkovich U. Special Photonic Crystal Fibers. Saarbrücken: Lambert Academic Publishing; 2011. 74 p. ISBN: 978-3-8465-3622-3

[34] Hoo YL, Jin W, Shi C, Ho HL, Wang DN, Ruan SC. Design and modeling of a photonic crystal fiber gas sensor. Applied Optics. 2003;**42**:3509-3515. DOI: 10.1364/AO.42.003509

[35] Ritari T, Tuominen J, Ludvigsen H, Petersen JC, Sorensen T, Hansen TP, et al. Gas sensing using air-guiding photonic bandgap fibers. Optics Express. 2004;**12**:4080-4087. DOI: 10.1364/OPEX.12.004080

[36] Pickrell G, Peng W, Wang A. Random-hole optical fiber evanescent-wave gas sensing. Optics Letters. 2004;**29**:1476-1478. DOI: 10.1364/OL.29.001476

[37] Fini JM. Microstructure fibres for optical sensing in gases and liquids. Measurement Science and Technology. 2004;**15**:1120-1128. DOI: 10.1088/0957-0233/15/6/011

[38] Matejec V, Mrazek J, Hayer M, Kask I, Peterka P, Kanka J, et al. Microstructure fibers for gas detection. Materials Science and Engineering: C. 2006;**26**:317-321. DOI: 10.1016/J.msec.2005.10.43

[39] Silva SF, Coelho L, Frazão O, Santos JL, Malcata FX. A review of palladium-based fiber-optic sensors for molecular hydrogen detection. IEEE Sensors Journal. 2012;**12**:93-102. DOI: 10.1109/JSEN.2011.2138130

[40] Huang P-C, Chen Y-P, Zhang G, Song H, Liu Y. Note: Durability analysis of optical fiber hydrogen sensor based on Pd-Y alloy film. The Review of Scientific Instruments. 2016;**87**:026104. DOI: 10.1063/1.4941749

[41] Zhang Y-N, Wu Q, Peng H, Zhao Y. Photonic crystal fiber modal interferometer with Pd/WO_3 coating for real-time monitoring of dissolved hydrogen concentration in transformer oil. The Review of Scientific Instruments. 2016;**87**:125002. DOI: 10.1063/1.4971321

[42] Minkovich VP, Monzon-Hernandez D, Villatoro J, Badenes G. Microstructured optical fiber coated with thin films for gas and chemical sensing. Optics Express. 2006;**14**:8413-8418. DOI: 10.1364/OE.14.008413

[43] Minkovich VP, Villatoro J, Sotsky AB. Tapered photonic crystal fibers coated with ultra-thin films for highly sensitive bio-chemical sensing. In: Proceedings of the European Optical Society Biennal Meeting (EOSAM); 8-12 October 2018. Netherlands: Delft; 2018. pp. 1-2

[44] Butler MA. Optical fiber hydrogen sensor. Applied Physics Letters. 1984;**45**:1007-1009. DOI: 10.1063/1.95060

[45] Bearzotti A, Caliendo C, Verona E, D'Amico A. Integrated optic sensor for the detection of H_2 concentrations. Sensors and Actuators B: Chemical. 1992;**7**:685-688. DOI: 10.1016/0925-4005(92)80386-C

[46] Wang C, Mandelis A, Garcia JA. Detectivity comparison between thin-film Pd/PVDF photopyroelectric interferometric and optical reflectance

hydrogen sensors. The Review of Scientific Instruments. 1999;**70**: 4370-4376. DOI: 10.1063/1.1150082

[47] Villatoro J, Diez A, Cruz JL, Andres MV. In-line highly sensitive hydrogen sensors based on palladium-coated single-mode tapered fibers. IEEE Sensors Journal. 2003;**3**:533-537. DOI: 10.1109/JSEN.2003.815789

[48] Ortigosa-Blanch A, Diez A, Gonzáles-Segura A, Cruz JL, Andres MV. Wavelength-codified fiber laser hydrogen detector. Applied Physics Letters. 2005;**87**:201104. DOI: 10.1063/1.2130531

[49] Zhao Z, Carpenter MA, Xia H, Welch D. All-optical hydrogen sensor based on a high alloy content palladium thin film. Sensors and Actuators B: Chemical. 2006;**113**:532-538. DOI: 10.1016/j.snb.2005.03.070

[50] Lin H, Gao T, Fantini J, Sailor MJ. A porous silicon-palladium composite film for optical interferometric sensing of hydrogen. Langmuir. 2004;**20**: 5104-5108. DOI: 10.1021/la049741u

[51] Xu T, Zach MP, Xiao ZL, Rosenmann D, Welp U, Kwok WK, et al. Self-assembled monolayer-enhanced hydrogen sensing with ultrathin palladium films. Applied Physics Letters. 2005;**86**:203104. DOI: 10.1063/1.1929075

[52] Juan Hu DJ, Lim JL, Park MK, Kao LT-H, Wang Y, Wei H, et al. Photonic crystal fiber-based interferometric biosensor for streptavidin and biotin detection. IEEE Journal of Selected Topics in Quantum Electronics. 2012;**18**:1293-1297. DOI: 10.1109/JSTQE.2011.2169492

[53] Li C, Qiu S-J, Chen Y, Xu F, Lu Y-Q. Ultra-sensitive refractive index sensor with slightly tapered photonic crystal fiber. IEEE Photonics Technology Letters. 2012;**24**:1771-1774. DOI: 10.1109/LPT.2012.2214379

[54] Nagel T, Ehrentreich-Forster E, Singh M, Shmitt K, Brandenburg A, Berka A, et al. Direct detection of tuberculosis infection in blood serum using three optical label-free approaches. Sensors and Actuators B: Chemical. 2008;**129**:934-940. DOI: 10.1016/j.snb.2007.10.009

[55] Yu W, Lang T, Bian J, Kong W. Label-free fiber optic biosensor based on thin-core modal interferometer. Sensors and Actuators B: Chemical. 2016;**228**:322-329. DOI: 10.1016/j.snb.2016.01.029

[56] Betancur-Ochoa JE, Minkovich VP, Montagut-Ferizzola YJ. Special photonic crystal modal interferometer for highly sensitive biosensing. Journal of Lightwave Technology. 2017;**35**:4747-4751. DOI: 10.1109/JLT.2017.2761738

[57] Betancur-Ochoa JE, Minkovich VP, Montagut-Ferizzola YJ, Minkovich PV. Highly sensitive biosensing based on a photonic crystal fiber modal interferometer. In: Proceedings of OSA International Conference on Optical Fiber Sensors (OFS 26); 24-28 September 2018; Lausanne: OSA; 2018. p. WF58(1-4)

[58] Yadav TK, Narayasnaswamy R, Abu Bacar MH, Mustafa Kamil Y, Mahdi MA. Single-mode tapered fiber-optic interferometric based refractive index sensor and its application to protein sensing. Optics Express. 2014;**22**:22802-22807. DOI: 10.1364/OE.22.022802

[59] Li X, Liu L, Zhao J, Tan J. Optical properties of sodium chloride solution within spectral range from 300 to 2500 nm at room temperature. Applied Spectroscopy. 2015;**69**:635-640. DOI: 10.1366/14-07769R

[60] Jiang Y. Fourier transform white-light interferometry for the measurement of fiber-optic extrinsic Fabry-Perot interferometric sensors. IEEE Photonics Technology Letters. 2008;**20**:75-77. DOI: 10.1109/LPT.2007.912567

Chapter 6

Dynamic Speckle Interferometry of Technical and Biological Objects

Alexander Vladimirov

Abstract

The theory of speckle dynamics in the image plane of a reflecting and thin transparent object is considered. It was assumed that the optical paths of the reflected and probing transparent object waves vary due to (1) translational motion, (2) oscillations with a period T, and (3) random relative displacements of pairs of scattering centers Δu (reflecting object) and random changes in the refractive index Δn (transparent object). The formulas relating the mean value, dispersion, and relaxation time of Δu and Δn values with the time-averaged radiation intensity at the observation point and the time autocorrelation function of this intensity are obtained. It is shown that at the averaging time multiple of T, the technique in real time allows to determine plastic deformations of the order of 10^{-3} on bases of the order of 10 microns, which is suitable for the control of elastic deformations on bases of the order of 100 microns. The possibilities of the method of averaged speckle images for the study of (1) features of the nucleation, start, and movement of the fatigue crack, and (2) the activity of living cells infected and not infected with the virus are demonstrated.

Keywords: theory, experiment, high cycle fatigue, live cell, speckle dynamics, estimation of time to failure, crack nucleation

1. Introduction

If a rough object is illuminated by coherent radiation, macroscopically homogeneous but microscopically inhomogeneous distribution of scattered radiation intensity emerges at some distance from the object and behind the lens that forms the object image. As the surface microrelief heights are random, the waves reflected from various microscopic areas of the surface have random amplitudes and phases. Mutual interference of these waves results in spotted or "speckle" structure of scattered radiation. When the object surface varies for some reason, the amplitudes and phases of the reflected waves change, so the speckle pattern also varies. Variation of wave amplitudes and phases can be caused by displacement of the object or its rigid rotation or by small variations in the distances between the area elements due to elastic or plastic strains of the surface. The reason may be a small or strong surface microrelief variation due to corrosion or loosening of the material with ionizing radiation, material fatigue, etc. The specified macroscopic or microscopic processes can occur simultaneously.

It is noteworthy that the speckle pattern can be formed when studying objects with a mirror surface by lighting them via mat glass. Phenomena occurring inside transparent objects can be studied in a similar way. Such objects are live cells cultured or precipitated on a transparent substrate. If there is an empty space, the cells can alter their shape and the intracellular processes randomly change the phases of the waves that have passed via the cell. So, by the character of the speckle pattern change, one can study and monitor the phenomena occurring on the surface or near the surface of reflecting objects or inside transparent media.

At present, one is familiar with application of variations, displacements, and interferences of speckle fields for the study of macroscopic phenomena, namely, for determination of transition, motion velocity, rotations, elastic or plastic strains of objects, gas, and fluid flows [1–7]. At the same time, the logic of speckle optics development and practical needs sets the task to analyze not only macroscopic but also microscopic processes occurring at the structural level. For example, a task like that appears in studies of the phenomena accompanying crack initiation in high-cycle fatigue of metallic materials as well as during the analysis of the processes occurring in the membranes and inside the cells of live systems. The rationale for such studies is related to development of a technique for assessing the remaining life of construction elements, and in case of biological objects, to individualized drug selection for a patient.

As the properties of materials are random at the structural level, the amplitudes and phases of the waves reflected from the object or that have passed through the object randomly vary in time and space. In the general case, the solution to the problem of establishing a relation between the parameters of wave phase dynamics and speckle dynamics is far from simple. Nevertheless, lately certain advances in solving such problems have been observed.

The objective of this publication is to familiarize the readers with the author's recent developments in the theory and application of dynamic speckle interferometry. They were aimed at study of the processes occurring in technical materials in their high-cycle fatigue and also in live cells subject to some external effects. The rationale for the specified studies, for the theory of the techniques proposed, the conducted experiments and their practical prospects are discussed in brief. The advantage of the techniques under discussion over the conventional speckle holographic techniques is the possibility for real-time study of reversible and irreversible processes. Most part of this chapter is devoted to dynamic speckle interferometry of high-cycle fatigue. When high-cycle fatigue is studied, the technique permits determination of the limiting local microrelief and surface shapes variations (deformations) with high sensitivity. When the variation rate of the named values is monitored, the time to the fatigue crack start can be determined. When the reaction of cells to viruses and bacteria is studied, an opportunity for timely development of procedures preventing and blocking their development progress appears.

2. High-cycle fatigue: the problems and the rationale for the studies

When technical objects are exploited, their various elements are affected by alternate forces. As load cycle number N in the materials increases, their properties change irreversibly resulting in fracture of the construction elements. In the literature, this phenomenon was called material fatigue. Depending on number N_a, of the cycles preceding the fracture, we distinguish low-cycle, high-cycle, and giga-cycle fatigue. Their borders are rather blurred. It is usually supposed that a part was destroyed in low-cycle fatigue if $N_a < 10^4$. If a part is exploited at $10^5 < \mathrm{N} < 10^7$, then high-cycle fatigue is supposed, and at $\mathrm{N} > 10^8$, the part is supposed to be exploited in giga-cycle fatigue.

At present, various features of low-cycle fatigue are fairly well characterized [8, 9]. It is explained by the fact that in low-cycle fatigue, sufficient plastic deformations of the objects normally appear. That is why conventional approaches developed for devices exploited under the conditions of quasi-static loading are applicable for monitoring and estimating of the remaining object life.

The situation with high-cycle as well as giga-cycle fatigue is different. According to various literature sources, from 50 to 80% of equipment is destroyed due to high-cycle fatigue [10–12]. Despite research history [13–16] and a great number of publications [17, 18], at present, there are no techniques for assessment and estimation of the remaining life of construction elements in their high-cycle fatigue that would meet the requirements of the engineering practice [11, 18]. According to [11, 18, 19], the situation has arisen due to absence of physical models of high-cycle fatigue. In the author's opinion, this circumstance is in turn related to absence of physical monitoring techniques that would permit to record the features of local fatigue damage accumulation without interruption of exploitation or fatigue testing of various objects. Analysis of studies [20–25] devoted to development of such methods shows that their practical application implies methodological issues.

Immediately after the creation of lasers and detection of spotted or speckle structure of scattered radiation, speckles were used to study fatigue phenomena [26–28]. However, because of high manpower effort and no monotone character of the recorded signal variation, the technique was not put to wide use.

The author learnt the difficulties of fatigue phenomena studies using conventional nondestructive testing techniques at first hand. Early in our research aimed at the study of high-cycle material fatigue, we used various modifications of optical, X-ray, magnetic, electric, and acoustic techniques (13 in all). The results obtained using these techniques were negative. The parameters of the signals either did not vary or the variations were at the same level as the hardware noise. As in high-cycle fatigue localization of fatigue damage takes place [11], it was supposed that the obtained negative result is related to a large base (averaging region) of the applied techniques. Hence, for measurements on a small base, we upgraded a speckle technique that was previously successfully used to monitor the damage in quasi-static deformation of specimens up to their fracture [29, 30].

Section 3 presents the results of the theoretical studies aimed at developing an optical technique intended to study the irreversible processes emerging while testing specimens for high-cycle fatigue. Section 4 discusses the experiments with high-cycle fatigue conducted using this technique.

3. Theory

This section discusses the results of theoretical underpinning of speckle techniques permitting the study of fatigue accumulation in periodical deformations of the objects. Sections 3.1 and 3.2 discuss the regularities of speckle dynamics observed while capturing the frames of the speckle image on fixed position of the object oscillations. Section 3.3 is dedicated to the theory of time-averaged speckle images.

3.1 Dynamic speckle interferometry of microscopic processes: reflecting object

Let the source of coherent radiation with wavelength λ located at point $\vec{s}$ illuminate the point scattering centers located in some region S in plane (x,y), as is shown in **Figure 1**.

A thin lens with focal distance f and diaphragm diameter D located in plane (η_x,η_z) forms the object image in plane (q_x,q_y). We regard all the waves at hand as linearly polarized in the same direction. We accept that phase φ_j of complex amplitude a_j of a wave scattered by the j-th center is random, and that the waves from all the scattering

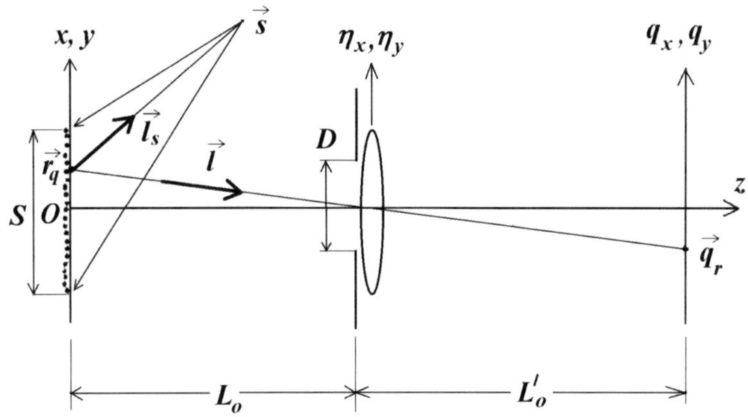

Figure 1.
Optical system for reflecting objects and designations.

centers come to an arbitrary point of plane (η_x,η_y). First, let us obtain an expression for radiation intensity $I\left(\vec{q}\right)$ at some point $\vec{q}$ of the object image plane. For total complex amplitude $A\left(\vec{\eta}\right)$ at arbitrary point $\vec{\eta}$ of plane (η_x,η_y), we have:

$$A\left(\vec{\eta}\right) = \sum_{j=1}^{N} a_j, \tag{1}$$

where N is the number of scattering centers. We will get the complex amplitude of light $A\left(\vec{q}\right)$ at point $\vec{q}$ adding up the amplitudes of the waves that came from the points of plane (η_x,η_y) to point $\vec{q}$, considering amplitude $P\left(\vec{\eta}\right)$ and phase $\exp\left[i\left|\vec{\eta}\right|^2/(2f)\right]$ transmission of the lens [3]:

$$A\left(\vec{q}\right) = \int_{-\infty}^{+\infty}\int P\left(\vec{\eta}\right) e^{\frac{i\left|\vec{\eta}\right|^2}{2f}} e^{ik\left|\vec{L}_q\left(\vec{\eta}\right)\right|} \sum_{j=1}^{N} a_j \, d\eta_x d\eta_y, \tag{2}$$

where i is an imaginary unit and $\vec{L}_q\left(\vec{\eta}\right)$ is a vector targeted from point $\vec{\eta}$ to point $\vec{q}$.

Let us take the relation between complex amplitude of light in proximity of point $\vec{r}_j$ and at point $\vec{\eta}$ in the same form as in [3]:

$$a_j\left(\vec{\eta}\right) = \sqrt{I_0\left(\vec{r}_j\right)}\,\xi\left(\vec{r}_j\right) e^{i\left\{k\left[\left|\vec{L}_s\left(\vec{r}_j+\vec{u}_j\right)\right|+\left|\vec{L}_\eta\left(\vec{r}_j+\vec{u}_j\right)\right|\right]+\phi_j\right\}}, \tag{3}$$

where $I_0 = I_0\left(\vec{r}\right)$ is the distribution intensity of the illuminating radiation, $\xi = \xi\left(\vec{r}\right)$ in the general case is the complex reflection coefficient that takes into account the proportion of the radiation going from point $\vec{r}$ to point $\vec{\eta}$, $\vec{L}_s\left(\vec{r}\right)$ is the vector targeted from point $\vec{r}$ to point $\vec{s}$, $\vec{L}_\eta\left(\vec{r}\right)$ is the vector targeted from point $\vec{r}$ to point $\vec{\eta}$, $\vec{u}_j$ is the vector of small displacement of the j-th center.

Let us take arbitrary point $\vec{q} = \vec{q}_r$ and its conjugated point $\vec{r}_q$. It is known that a wave going from point $\vec{r}_q$ forms an Airy pattern as the result of light diffraction on a diaphragm with diameter D and the center at point $\vec{q}_r$. The radius of central spot b_s of the pattern equals 1, $22\lambda L_0'/D$, where L_0' is the distance from the lens to plane (q_x, q_y). Region with radius $a_s = b_s/m$, where m is the magnification of the lens, corresponds to regions with radius b_s in plane (x,y). It is known that 85% of the energy of the wave that passed via the lens falls on the central spot of the Airy pattern. We will neglect the energy that falls on the sections beyond the region with radius b_s. This in turn means that we suggest that the waves from the scattering centers located only in the region with radius a_s with the center at point $\vec{r}_q$ come to point $\vec{q}_r$. Now, let N be the number of such centers.

Let us further assume that the region with radius a_s, values D, and $\left|\vec{u}\right|$ is small compared to the distances from the object to the radiation source and to the lens as well as that from the lens to the image plane. Regarding expression $\left|\vec{L_s}\right| = \left|\vec{L_s}\left(\vec{r} + \vec{u}\right)\right|$ as continuous function $\vec{r}$ and $\vec{u}$, let us expand it in a Taylor series in a small neighborhood of point $\vec{r} = \vec{r}_q$, $\vec{u} = 0$, taking into account the derivatives to the first-order inclusive. Let us similarly expand expression $\left|\vec{L_\eta}\left(\vec{r} + \vec{u}\right)\right|$ in the vicinity of points $\vec{r} = \vec{r_q}$, $\vec{\eta} = 0$, and $\vec{u} = 0$, and expression $\left|\vec{L_\eta}\left(\vec{q}\right)\right|$ in the vicinity of points $\vec{\eta} = 0$ and $\vec{q} = \vec{q}_r$. One can familiarize oneself with the results in specified expansions in [31, 32]. Taking the obtained expansion, as well as formulas (3) and (2) into account, we in [31] obtained an expression for $A\left(\vec{q}\right)$ and $I\left(\vec{q}\right)$:

$$A\left(\vec{q}\right) = \sqrt{I_{01}}e^{i\psi_j}\sum_{j=1}^{N} e^{i\left\{k\left[\vec{u_j}\left(\vec{l_s} + \vec{l}\right)\right] + \theta_j\right\}}, \tag{4}$$

$$I\left(\vec{q}\right) = A\left(\vec{q}\right) \times A^*\left(\vec{q}\right) = I_{01}N + 2I_{01}\sum_{\kappa=1}^{K} \cos\left[\Delta\phi_\kappa + \Delta\theta_\kappa\right], \tag{5}$$

where $\vec{l_s} = \vec{l_s}\left(l_{sx}, l_{sy}, l_{sz}\right)$ and $\vec{l} = \vec{l}\left(l_x, l_y, l_z\right)$ are unit vectors targeted from point $\vec{r}_q$ to the radiation source and to the observer, respectively; $\theta_j = k\vec{r_j}\left(\vec{l_s} + \vec{l}\right) + \varphi_j$, complex amplitude $\sqrt{I_{01}}e^{i\psi_j}$ determines the expression preceding the summation sign; $\Delta\phi_\kappa = k\Delta u_\kappa = k\Delta\vec{u_\kappa}\left(\vec{l_s} + \vec{l}\right)$, $\Delta\vec{u_\kappa}$ is the relative displacement vector of the κ-th pair of scattering centers; and $\Delta\theta_\kappa = \theta_j - \theta_m, j \neq m$, $\kappa = 1, 2 \ldots K$, $K = N(N-1)/2$.

Then, let us assume that process $\Delta u_\kappa(t)$ is random, and for a fixed κ, there is correlation of values Δu_κ in time. Assuming that all the random values are independent, in [31], we obtained an expression for temporal normalized autocorrelation function of radiation intensity $\eta_{1,2}(t_1, t_2)$ at point $\vec{q}$:

$$\eta_{1,2}(t_1, t_2) = \cos\left[\langle x_2\rangle - \langle x_1\rangle\right] \times e^{-\frac{1}{2}k_{11} - \frac{1}{2}k_{22} + k_{12}}, \tag{6}$$

where $\langle x_1\rangle$ and $\langle x_2\rangle$ are values $\Delta\phi_\kappa$ average by the ensemble of objects at time points t_1 and t_2, respectively, k_{11} and k_{22} are variances of values $\Delta\phi_\kappa$ at time points t_1 and t_2, respectively, and k_{12} is the correlation coefficient of phase differences $\Delta\phi_\kappa$ at time points t_1 and t_2.

If process $\Delta u_\kappa(t)$ is stationary, then $\langle x_1 \rangle = \langle x_2 \rangle$ and $k_{11} = k_{22}$, and instead of (6), we have:

$$\eta(\tau) = e^{-k_{11} + k_{11}\rho_{12}(\tau)}, \tag{7}$$

where $\rho_{12}(\tau)$ is a normalized autocorrelation function of relative displacements, $\tau = t_2 - t_1$. From (7), it follows that if τ exceeds correlation time τ_0 of values Δu_κ, function $\eta(\tau)$ levels off to $\eta^* = \exp(-k_{11})$. In our experiments discussed below, this circumstance was used to determine value k_{11} as the parameter of fatigue damage to the specimens.

At values Δu_κ that are small compared with λ, it is convenient to exclude the permanent part of $\eta = \eta^*$, proceeding to a new normalized autocorrelation function $\eta'(\tau) = [\eta(\tau) - \eta^*]/[\eta(0) - \eta^*]$. It is easy to show that at $\Delta u_\kappa \ll \lambda$, we have $\eta'(\tau) \cong \rho_{12}(\tau)$. Therefore, the corresponding normalized spectral functions $g_I(\omega)$ of intensity fluctuation and relative displacements $g_{\Delta u}(\omega)$ are equal. The correlation (relaxation) times τ_0 of values Δu_κ and fluctuations of radiation intensity τ_κ in the conjugated region are also equal.

At $\Delta u_\kappa \geq \lambda$, establishing a relation between $g_{\Delta u}(\omega)$ and $g_I(\omega)$ is not a simple task. In [33], it was done for the case when function $\rho_{12}(\tau)$ is a Gaussian function. It was shown that in this case, $g_{\Delta u}(\omega)$ and $g_I(\omega)$ are also Gaussian functions, with the function range $g_I(\omega)$ at level 1/e is k_{11} times as wide as that of function $g_{\Delta u}(\omega)$.

3.2 Theory of speckle dynamics of microscopic processes: transparent object

Let us discuss an optical system (**Figure 2**) that forms an image of a thin transparent (phase) object. Let the source of coherent light 1 illuminate thin diffuser 2 consisting of point stationary diffusers chaotically located near plane (x,y). The waves spreading from point centers first pass through various sections of thin transparent object 3 and then via lens with diagram 4. Planes (x,y) and (q_x, q_y) are conjugated. Let us accept that the sum of the diffuser thickness, the distance from the diffuser to the object, and the object thickness are less than the longitudinal resolution of the lens. To simplify the calculations, let us suggest that the refraction indexes of the medium inside and outside the diffuser are same and equal 1. Let us assume that the optical paths of the sounding waves in the object 3 randomly change, in time, and let us neglect the refraction of waves within the object. Let us take the conjugated points 6 and 7. Reasoning the same way as we did in Section 3.1,

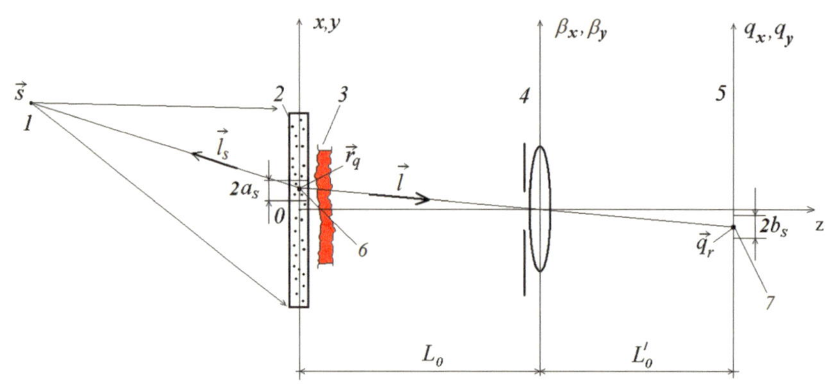

Figure 2.
Optical system for a transparent object: 1—light source, 2—diffuser, 3—transparent object, 4—lens with diaphragm, 5—image plane, and 6 and 7—conjugated points.

we can obtain an expression for radiation intensity $I\left(\vec{q}\right)$ at point 7, and also for normalized autocorrelation function $\eta(t_1, t_2)$ of this intensity. Such calculations were performed in [33]. It was obtained that the expression for $\eta(t_1, t_2)$ is precisely formula (6), but the value Δu_κ is determined by a different formula:

$$\Delta u_\kappa = u_j - u_m \tag{8}$$

where κ is the wave pair number again, and

$$u_j = \int\limits_{l_j} [n(l) - n_0] dl, \; u_m = \int\limits_{l_m} [n(l) - n_0] dl, \tag{9}$$

are the optical path lengths of the j-th and m-th sounding waves, respectively; l_j and l_m are the paths of the corresponding waves, and $n(l)$ is the distribution of the refraction index along the wave path. The above features of functions $\eta(t_1, t_2)$ and $g_{\Delta u}(\omega)$ remain valid.

3.3 Theory of time averaging method for speckle images

The results of the theory discussed above in Sections 3.1 and 3.2 were used in studies of irreversible processes arising in metals with their high-cycle fatigue [31, 34] and in live cells [35]. In the cases when spatially homogeneous and temporally stationary random variation of optical wave paths occurred at the structural level, good coincidence between the theory and the experiment was observed.

However, the drawback of the theories discussed in Sections 3.1 and 3.2 is the difficulty of application in the cases when random variations of the wave phases occur due to various simultaneous processes occurring at different rates.

For example, in high-cycle metal fatigue, the phase of the wave reflected from part of the surface can vary as a result of translational motion of the object, its elastic or plastic deformation, phase transformation, and formation of microcracks.

When a live cell is studied, the wave phase can vary due to diffusion of the substances through the membrane, endocytosis (capture of large particles due to local cell shape variation), and protein synthesis as well as cell motion.

To overcome the drawback named in [32, 36, 37], the theory was upgraded. The idea consisted in application of the time averaging procedure to speckle dynamics. If the characteristic time τ_0 of wave phase variation corresponding to the fastest process is known, then the averaging time T_1 of the recorded optical signals can be taken larger than τ_0. In this case, speckle dynamics will be formed due to occurrence of slower processes; moreover, interpretation of the experimental data can get simplified.

Let us discuss the results of the theoretical studies, first for a reflecting object, and then for a transparent object. In [32], from the model of scattering reflecting object discussed in Section 3.1, a problem concerning speckle dynamics in the image plane of a flat surface performs a complex motion. It was supposed that the scattering centers located in plane (*xoy*) (**Figure 1**) simultaneously (1) move at a low rate toward *ox* axis, (2) perform periodic motions with the same period and amplitude toward *ox* axis, and (3) randomly move in space. It was supposed that the difference of irreversible random center displacements $\Delta u_\kappa = \Delta \vec{u_\kappa}\left(\vec{l_s} + \vec{l}\right)$ appear after another cycle of the object oscillation. The averaging time must be equal to the oscillation period or be divisible by it. For the time-average radiation intensity $\tilde{I}\left(\vec{q}\right)$ at arbitrary point $\vec{q}$ and for normalized temporal autocorrelation function $\eta(t_1, t_2)$ of this intensity, the following was obtained:

$$\tilde{I} = I_1 + I_2 e^{-\sigma^2/2} \cos\left(x + \alpha\right), \tag{10}$$

$$\eta(t_1, t_2) = \eta(u_x) \times e^{-k_{11}/2 - k_{22}/2 + k_{12}(t_1, t_2)} \times \cos\left(\langle x_1 \rangle - \langle x_2 \rangle\right), \tag{11}$$

where I_1, I_2, α are constants, x and σ^2 are the mean value and variance of Δu_κ, respectively, obtained by averaging over the time T_1 and over the region with a radius $\Delta x = 2a_s$. This area is located in the vicinity of the point that conjugates the point $\vec{q}$. Function $\eta(u_x)$ is a normalized temporal autocorrelation function corresponding to a translational displacement of object toward ox axis; u_x is value of displacement. Values k_{11}, k_{22}, k_{12}, $\langle x_1 \rangle$, $\langle x_2 \rangle$ have the same meaning as in formula (6). However, now they are parameters characterizing random wave pair phase differences averaged by time T_1.

A similar problem concerning speckle dynamics in the image plane of a transparent object was solved in [37]. An optical system presented in **Figure 2** was discussed. It was supposed that object 3, located near thin diffuser 2, is a thin transparent plate whose lateral surfaces are parallel to plane (x,y). As in the previous case, the discussion was about the complex motion of the plate consisting of (1) its translational low-rate motion along ox axis, (2) its periodic motion toward ox axis, and (3) random small variation of the optical thickness of the plate. The formula for time-average intensity coincided with formula (10), and the formula for the temporal autocorrelation function of this intensity was determined by the expression:

$$\eta(t_1, t_2) = \left[e^{-k_{11}/2 - k_{22}/2 + k_{12}(t_1, t_2)} \times e^{-\tilde{k}_{11}/2 - \tilde{k}_{22}/2 + \tilde{k}_{12}(t_1, t_2)}\right] \times \cos\left(\langle x_1 \rangle - \langle x_2 \rangle + \langle \tilde{x}_1 \rangle - \langle \tilde{x}_2 \rangle\right). \tag{12}$$

Now, in formula (12), the value x is a wave pair phase difference averaged by time and region with diameter Δx, whose changes are determined by variations of the optical thickness of the plate using formula (9). Parameters k_{11}, k_{22}, k_{12}, $\langle x_1 \rangle$, $\langle x_2 \rangle$ in formulas (11) and (12) have the same meaning. Parameters $\tilde{k}_{11}$, $\tilde{k}_{22}$, $\tilde{k}_{12}$, $\langle \tilde{x}_1 \rangle$, $\langle \tilde{x}_2 \rangle$ are similar to values k_{11}, k_{22}, k_{12}, $\langle x_1 \rangle$, $\langle x_2 \rangle$, but they are the parameters of the speckle dynamics emerging due to the roughness of the plate performing a translational motion.

If the roughness is homogeneous, i.e., $\tilde{k}_{11} = \tilde{k}_{22}$, $\langle \tilde{x}_1 \rangle = \langle \tilde{x}_2 \rangle$, then instead of (12), we have:

$$\eta(t_1, t_2) = \eta(u_x) \times e^{-k_{11}/2 - k_{22}/2 + k_{12}(t_1, t_2)} \times \cos\left(\langle x_1 \rangle - \langle x_2 \rangle\right), \tag{13}$$

where

$$\eta(u_x) = e^{-\tilde{k}_{11} + \tilde{k}_{11}\rho_{12}(u_x)} \tag{14}$$

is a normalized temporal autocorrelation function of radiation intensity corresponding to the translational motion of the plate, and $\rho_{12}(u_x)$ is a normalized temporal autocorrelation function of phase difference of wave pairs changing in time as a result of movement of a rough transparent plate.

Let us note that the theory of dynamics of time-averaged speckles in the image plane of a thin transparent object in the absence of its displacement and oscillation was discussed in [36].

3.4 Discussion of theoretical results

Therefore, if translational motion of a reflecting or transparent object is absent, the averaging time is divisible by the cyclic loading period, and there are no irreversible deformations in the object, then, according to Eqs. (9)–(11), the observer in the image plane will see a pattern of averaged speckles invariable in time. If irreversible processes that alter the optical paths of the waves emerge in a small region of the object at some phase of oscillation, then the speckle pattern in the conjugated region will change. As the formulas for normalized autocorrelation functions (6) and (11) at $\eta(u_x) = 1$ coincide, the pattern of averaged speckles can be regarded as a speckle pattern of some stationary object. In case of the emergence of irreversible processes that alter the shape of the reflecting object at the structural and/or macroscopic level, or the density of the transparent object, the speckle pattern of such a "stationary" object will vary.

For the reflecting object, value x in the cosine input is proportional to the displacement difference of surface points $\Delta u = \Delta \vec{u} \left(\vec{l_s} + \vec{l} \right)$ located at some characteristic (mean) distance Δx. If the deformation variance is small $(k_{11}, k_{22} \to 0)$, then values $\tilde{I}$ and η in the above formulas will depend on Δu by the law of cosines. Thus, the formulas go into the ratios known in the shear variants of holographic, correlation, and dynamic speckle interferometry. Let us note that there emerges a potentiality to determine the components of vector $\Delta \vec{u}$ by means of recording and procession of speckle images recorded simultaneously at different directions of the object illumination or observation.

4. Experiment

4.1 Dynamic speckle interferometry of flat specimens in periodic bending

In our first experiment, the results of theoretical analysis presented in Section 3.1 were applied to study fatigue phenomena emerging in high-cycle fatigue of medium-carbon steel 50 [34]. The scheme of specimen loading is presented in **Figure 3**; the dimensions and shapes of the specimen as well as the speckle image of the control area are shown in **Figure 4**. Before testing, the sample was subjected to fine grinding and annealing. A 2 mm thick flat specimen was loaded with 50 Hz frequency; the number of cycles reached 1,200,000; and maximum cycle amplitude σ_{max} varied from 0.2 to 0.82 σ_{02}, where σ_{02} is the flow limit of steel 50. The surface area near the maximum stresses was illuminated by a laser module with wavelength λ = 655 nm and 20-mW power. Speckle images with magnification m = 0.1 were captured at a certain phase of the object oscillation and entered into a computer with a frequency about 10 Hz. The minimum speckle size $2b_s$ in the object image plane equaled 40 μm. Films of 20- to 60-s duration were recorded at various stages of specimen testing. To determine value η that is involved in formula (7), we took digital values of intensity at one point (pixel), but at different time points t_1 and $t_1 + \tau$. The digital value of intensity $I_1(t_1)$ corresponded to the beginning of the film, and $I_2(t_1 + \tau)$ corresponded to a frame at time point $(t_1 + \tau)$. A part of the surface size $2a_s = 2b_s/\mathrm{m}$ was regarded as one "object" of the object ensemble. It was supposed that this "object" was located in the vicinity of a point conjugated to the pixel we are discussing. The array of the areas' size $2a_s$ located in surface area size 1×4 mm (1 and 4 mm along axes x and y, respectively) was regarded as an object ensemble. Measurement of η was determined by formula:

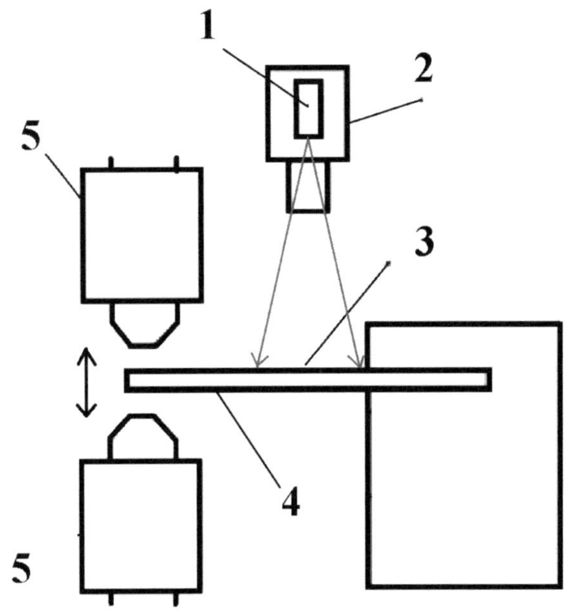

Figure 3.
Scheme of periodic bending: 1—laser module, 2—TV camera, 3—control zone, 4—specimen, and 5—electromagnet.

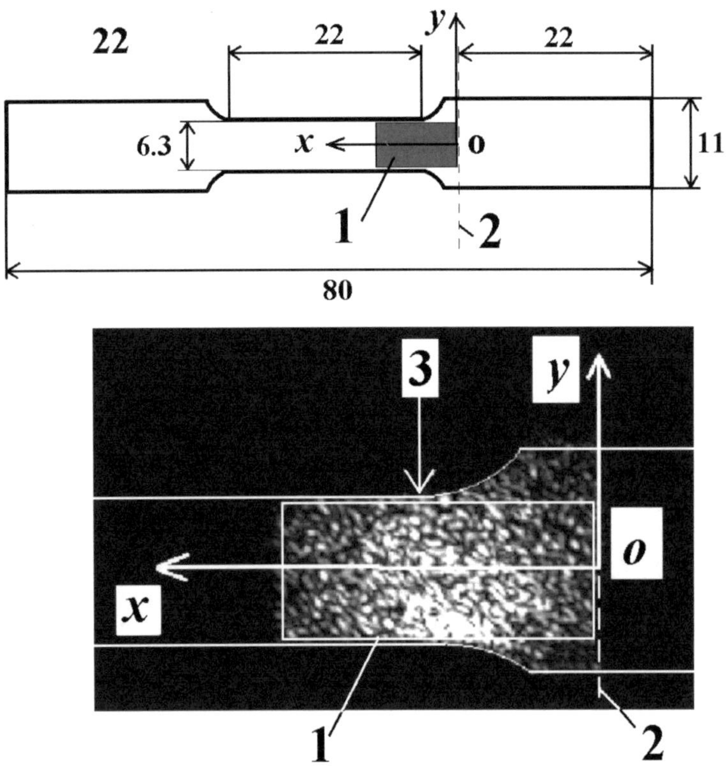

Figure 4.
Dimensions and shape of the specimen (top) and speckle pattern (bottom): 1—control zone, 2—clamp line, and 3—maximum stress spot.

$$\eta(\tau) = \frac{\langle [I_1(t_1) - \langle I_1(t_1)\rangle][I_2 - \langle I_2(t_1+\tau)\rangle]\rangle}{\sigma_1 \times \sigma_2} \quad (15)$$

where the angle brackets denote arithmetic mean values in the named object ensemble, and σ_1 and σ_2 are standard deviations of values $I_1(t_1)$ and $I_2(t_1 + \tau)$.

Figure 5 presents a typical dependencies $\eta = \eta(\tau)$ obtained early in the fatigue testing experiment at $\sigma_{max} = 0.4\sigma_{02}$. The dark spots denote the experiment; and the dashed line denotes a Gaussian autocorrelation function. The deviation of the experimental data from the theory is in the 5% range. As is seen from the graphs, the autocorrelation function declines from 1 to the fixed level $\eta^* = 0.83$. According to our theory, the decline of value η and leveling-off are the indicators of homogeneity and stationarity of random process $\Delta u = \Delta u(t)$ discussed in Section 3.1. As the directions of illumination and observation are close to the surface normal, $\Delta u = \Delta \vec{u}\left(\vec{l_s} + \vec{l}\right) \cong 2\Delta u_z$, where Δu_z is the projection of vector $\Delta\vec{u}$ on the surface normal. Using the surface decline on 1/e to level η^*, we obtain that the intensity fluctuation correlation time is 7 s. One can accept that this time rivals with relaxation time τ_0 of random value Δu_z, or difference of the displacement of two surface points toward the normal. These points are located in a 400-μm area. From formula $\eta^* = \exp[-k_{11}] = \exp\left[-(2\pi/\lambda)^2 4\sigma^2_{\Delta u_z}\right]$, we obtain that standard deviation $\sigma_{\Delta u_z}$ of values Δu_z equals to 23 nm.

Analysis of the experimental data obtained in the sections of stationary surface relief variations showed the following. When σ_{max} varies from $0.2\sigma_{02}$ to $0.82\,\sigma_{02}$, correlation time τ_0 of relative displacements Δu_z varies from 60 s to values of the 10^{-1}-s order. At the same time, when σ_{max} varied in the named range, value $\sigma_{\Delta u_z}$ stayed practically invariable, at the 5-nm level on the average. Rare measurement lunges of $\sigma_{\Delta u_z}$ from 10 to 20 nm were observed.

In many cases, we recorded very complex dependences $\eta = \eta(\tau)$. Their analysis enabled us to suppose that they corresponded to several processes occurring simultaneously and altering the phases of scattered waves at different rates. Because of difficulties in the interpretation of such data, the technique was upgraded. A theory of time-averaged speckles discussed in Section 3.3 was proposed. Applications of the results of this theory are discussed in the next section.

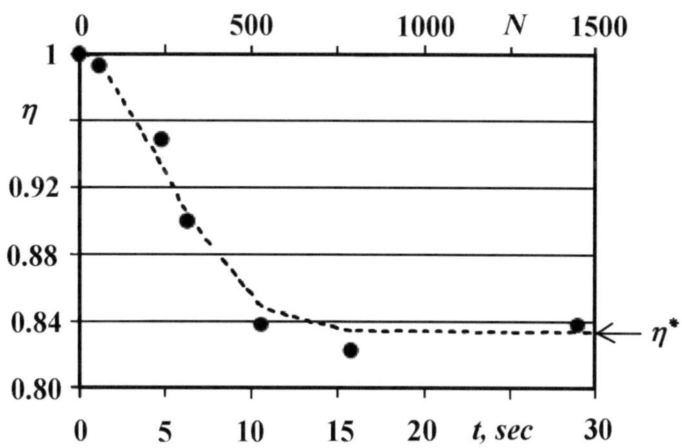

Figure 5.
Temporary autocorrelation function of intensity fluctuations due to metal fatigue: •—experiment and ---- theory.

4.2 Application of time-averaged speckles

We used the technique of time-averaged speckles to study the micro- and macro-variations of the surface shape in high-cycle fatigue of constructive and model materials. The main objective of the conducted research work was to clear up the matter of what happens in the material during crack initiation.

Beams made from pipe steel 09Г2С loaded by three-point bending were used in the experiment. The shape and dimensions of the specimens as well as the position of the bumps are presented in **Figure 6**. A Charpy notch was made in the beam to localize the crack initiation spot. The surface of the beams was subject to fine grinding, and then they were annealed in vacuum. After annealing, the surface of some specimens was polished. After polishing, roughness parameter R_a was in the 1- to 50-nm range. Let us note that for both polished and unpolished specimens, we observed the same results in a qualitative sense.

The specimens were tested in a resonance-type machine of MIKROTRON (RUMUL) type with near-100 Hz frequency and a 0.1 load ratio. The value of the maximum force exerted (1.1 kN) was selected experimentally from the condition of emergence of a 0.1-mm long crack after hundreds of thousands of loading cycles. Variation of resonance frequency by 10% usually preceded the emergence of a crack of this length.

The scheme of the optical setup applied for recording of the average speckle images is presented in **Figure 7**. The optical setup was located on the platform of the testing machine. Object 4 was illuminated by beam 2 from laser module 1 of a KLM-H650-40-5 type with the wavelength of 0.65 μm and 40-mW power. As speckles do not emerge when a mirror surface is illuminated, mat glass 3 was put into the lighting beam to form speckle fields when specimens with the polished surface were tested. The speckle image was recorded in the specimen image plane. The magnification of the optical system equaled 0.7. The diaphragm size of lens 5 was selected so that the minimum speckle size slightly exceeded that of a photocell in the photocell array of the TV camera 6. A monochrome VIDEOSCAN-415M-USB TV camera with an array containing 782 × 582 photocells of 8.3 × 8.3 μm size was used in the experiments. Averaging time T_1 equaled 50 oscillation periods. We supposed that in the most damaged surface area, this time exceeded the correlation time of wave pair phase difference τ_0 which is equal to 0.1 s.

Figure 8 presents distribution of correlation coefficient η found by two 8-bit digital speckle images of the specimen.

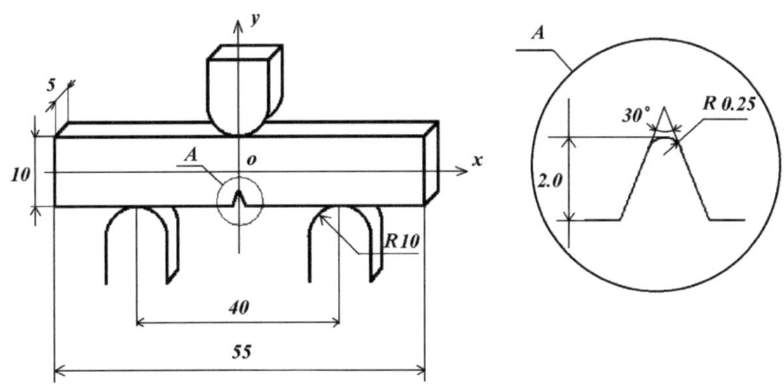

Figure 6.
Shape and dimensions of the specimen.

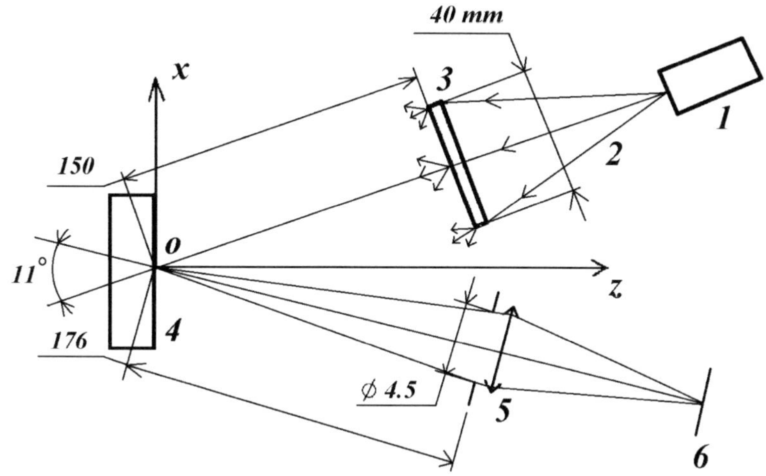

Figure 7.
Scheme of the optical setup. 1—laser module, 2—illuminating radiation, 3—mat glass, 4—specimen, 5—lens with diaphragm, and 6—photosensor array of the TV camera.

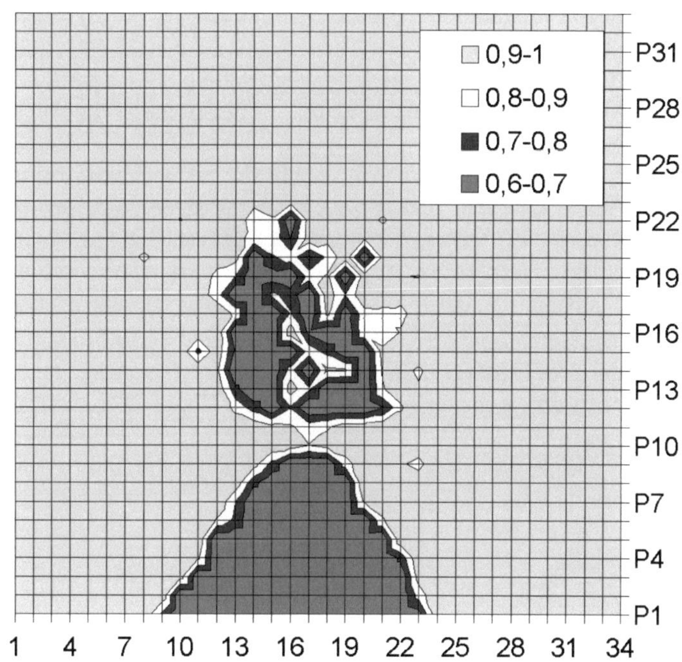

Figure 8.
Distribution of correlation coefficients η found in a 4 × 4-pixel section.

The first image was obtained at loading cycle number N_1 equal to 57,000, and the second image was obtained at cycle number N_2 = 92,000. Speckle field variations started at N_2 equal to 72,000 cycles. Value η was found using formula (15).

Figure 9 presents three-dimensional surface profiles near the notch at N_1 = 57,000 and N_2 = 92,000. The profiles were recorded using an optical profilometer WYKO NT-1100 with a 3-nm height measuring error.

As it is seen from **Figure 9**, at 92,000 cycles, two zones emerged in front of the notch. The first zone is a pitch about 0.5 μm deep with the diameter of 500 μm. The pitch center was at the distance of about 250 μm from the notch tip. Besides, a second small zone of about 50 × 100-μm size where a fatigue crack initiated

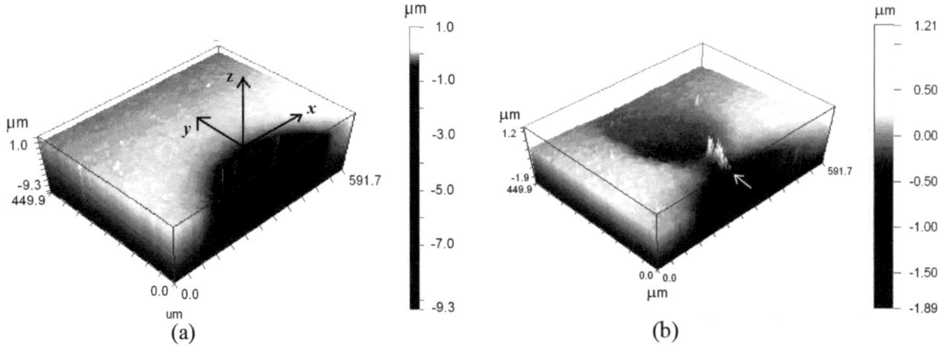

Figure 9.
Surface profiles: (a) at 57,000 cycles and (b) at 92,000 cycles.

emerged immediately at the notch tip. This area is shown by an arrow in **Figure 9b**. This zone consisted of irregularities of 5- to 1-μm transverse size and the height of scores of nanometers.

Figure 10 presents joint graphs showing altered relief heights Δh and correlation coefficient variation η in the pitch image plane along ox axis.

The scan line of the profilometer passed through the pit center, and value η was determined using formula (15) by scanning of a 4×4 pixel-size section. It is seen from the picture that the maximum variations of value η fall on the steepest surface slopes in the pit. Minimum variations of value η correspond to the sections beyond the pit and at its bottom.

Comparison of surface shape variation within the pit with variations of value η in the conjugated region showed that these variations agree with formula (11). The mean value of the scattering center displacement difference at the bottom of the pit can be accepted as zero. In this case, in formula (11), variation of value η will be determined by the parameters contained in the exponent. Values $[\lambda/(4\pi)]\sqrt{-\ln(\eta)}$ corresponding to the bottom of the pitch equaled several nanometers which corresponded to the profilometer data. In **Figure 11**, for five scan lines, black dots show dependence of value η that falls on the steepest slopes of the surface, from the

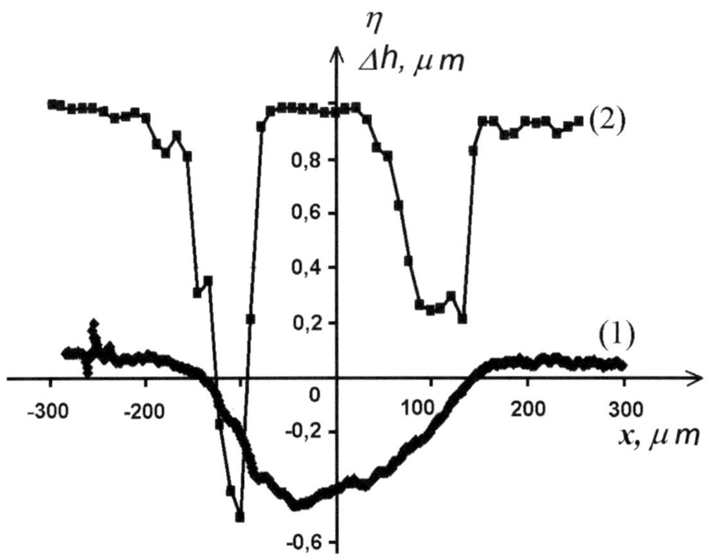

Figure 10.
Joint dependences (1) $\Delta h(x)$ and (2) $\eta(x)$.

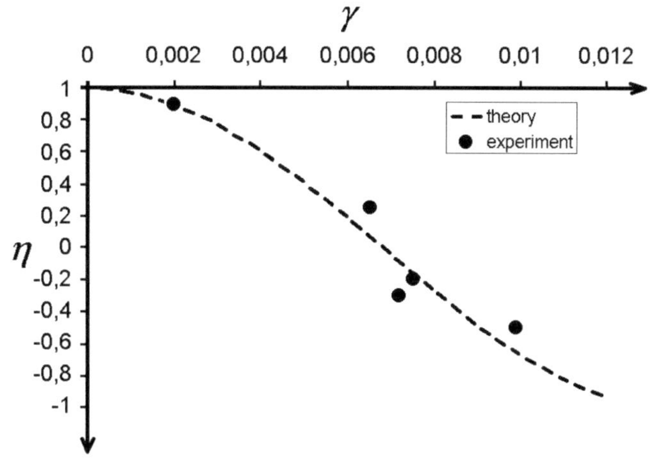

Figure 11.
Theoretical (—) and experimental (•) dependences $\eta(\gamma)$.

maximum measurement of the surface profile slope ratio γ. The latter value was found by the surface profile in the graph. The dotted line corresponded to formula (11), in which it was accepted that $\langle x_1 \rangle = 0$ and $\eta(u_x) = 1$; the exponent was substituted by 1 and the cosine argument was $(4\pi/\lambda)\gamma 2x_s$, where $2x_s$ is the size of the surface area equal to the linear resolution of the lens (12 μm). As seen from **Figure 11**, we have fairly good coincidence of the theory and experiment.

As seen from **Figure 9**, about half of the crack formation region lies within the pit, and the other half is beyond the pit. For the latter section, the measurement of η equaled 0.8. Suppose in (11) if the cosine equals 1, $\eta(u_x) = 1$, and $k_{11}, k_{12} = 0$, we obtain that the roughness parameter R_a variation equals 20 nm.

The measurement of R_a was determined by the equation $R_a \cong 0.8\lambda\sqrt{-2\ln\eta}/[2\pi(l_{sz} + l_z)]$, which is fair if the relief heights are distributed by the Gaussian law. Here, l_{sz} and l_z are the projections of unit vectors $\vec{l_s}$ and $\vec{l}$ on axis z, respectively, and $l_{sz} = l_z = 0,98$.

According to the profilometer data, variation of value R_a in the segment equaled 25 nm.

Figure 12 presents typical dependences of $\eta(N)$ that correspond to five surface sections of another specimen. The graph numbers coincide with the section numbers. The fatigue test conditions for this specimen did not differ from those for the previous one; the test was stopped only after the crack initiation and a small increase in its length.

Analysis of three-dimensional profiles of the surfaces of two specimens showed that they were the same in the qualitative sense. The surface spots for plotting of dependences $\eta(N)$ in **Figure 12** were selected so that they could clear up the emergence order of the two above zones at the notch tip. This information can be beneficial for creation of crack initiation physical models. Section 1 was selected beyond the contraction; sections 2 and 3 were selected on the path of the motion of the plasticity zones in front of the crack. Section 4 was located at the very notch tip, and section 5 was taken on the contraction edge on the line approximately passing through its center parallel to the specimen axis. The decrease of value η from 1 to 0.95–0.97 for section 1 at the end of the experiment is caused by the hardware noises and partially by slight translational motion due to degradation of the material. The centers of sections 2 and 3 rivaled, but their size differed. Section 2 of 22×22-μm size consisted of four regions of Δx = 11-μm size, and in section 3, there were 16 such regions. It is evident that dependences η (N) for these two sections

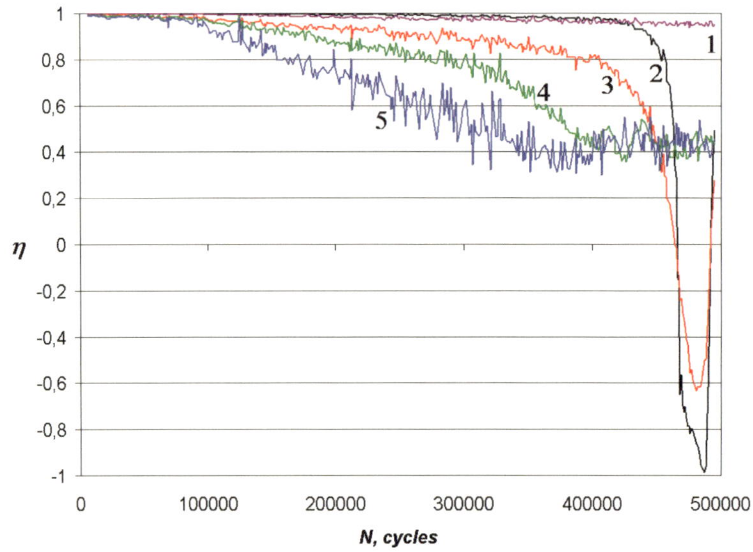

Figure 12.
Dependences η(N) corresponding to five sections of the surface: 1—section of 44 × 44-μm size at the distance of 640 μm from the notch tip. 2—section of 22 × 22 μm size at the distance of 180 μm, 3—section of 44 × 44-μm size at the distance of 180 μm, 4—section at the notch tip of 44 × 44-μm size, 5—section of 44 × 44-μm size on the slope of the contraction.

twice arrive at zero nearly simultaneously; however, their minimum measurements differ. For section 2, value a η nearly reaches -1, and for section 3 the minimum measurement equals -0.62. From the viewpoint of our theory, in compliance with formula (11) reaching -1 by value η means its dependence on $\langle \Delta u_z \rangle$ by the law of cosines with amplitude $\exp(-k_{22})$ equaling 1 ($k_{22} = 0$), and values $k \times 1.96 \times \Delta\gamma \times \Delta x/\lambda$ in every small region of 11-μm size reach value $\pi/2$. Let us take value $\Delta\gamma$ that characterizes a rigid surface rotation for the measure of plastic deformation on the 11-μm base. Then it follows from our analysis that at $\eta = -1$, a homogeneous plastic surface deformation equaling 7.5×10^{-3} occurs in section 2. For section 3 amplitude, $\exp(-k_{22})$ equals 0.62. Thus, we get that the plastic deformations in section 3 of 44 × 44-μm size are inhomogeneous, and the standard deviation of values $\Delta\gamma$ from 7.5×10^{-3} equals 3.3×10^{-3}.

It is seen in **Figure 12** that notable deviations of dependences 4 and 5 from the horizontal line start practically simultaneously at 70,000–80,000 cycles. This suggests that the formation of irregularities in the small zone and the contractions probably start simultaneously. Above, it was shown that a decline of value η from 1 to about 0.8 in the conjugated region corresponds to crack initiation at the notch tip. For section 4, such a decline occurred at about 280,000–300,000 thousand cycles. Then, dependence 4 levels off at $N \approx 400{,}000$. This suggests cessation of irreversible process in section 4 after the crack start and the increase in the crack length. A decrease of value η to 0.45 at 330,000 cycles approximately in linear fashion is characteristic for to section 5. Then dependence $\eta(N)$ also goes to the horizontal section which speaks for the completion of contraction formation.

4.3 Studying fatigue of Plexiglas

The experiments conducted with Plexiglas of "ACRUMA" brand aimed at studying the peculiarities of fatigue damage accumulation in the volume of the specimen shown in **Figure 6**. The specimens were made from a plate 5-mm thick. The protective film preserving the polished surfaces was taken off before the fatigue test. The equipment shown in **Figure 7** was used to record the speckle images. The specimen was illuminated through mat glass, the sounding waves passed through the specimen. The observation direction was selected either along the normal or at the angle of 30° to the normal. In the latter case, irreversible processes in the depth of the specimen could be controlled by distribution of value η. The three-point bending test under the condition of high-cycle fatigue was run on the machine discussed in the previous section. By now, an article has been prepared on the basis of the obtained data, which is now being reviewed by the editors of an optical magazine. As agreed with the magazine editors, the results of the conducted experiments will be available after publishing. We will only note that we were for the first time able to view in detail the process of transition of originally continuous three-dimensional medium into a destroyed state. Using the formulas in Section 3.3, we evaluated the limiting variations of the refraction index and density prior to fatigue destruction of Plexiglas. The crack initiation processes in steel and Plexiglas were very similar, even coinciding in some details. The difference consists in the fact that metals have the mechanism of energy dissipation connected to dislocation motion. That is why during local plastic deformation, variation of the object shape (emergence of a contraction) can take place. An amorphous material like Plexiglas does not have such a mechanism. That is why plastic deformations come into effect as volume/density variations.

4.4 Dynamic speckle interferometry of intracellular processes

In our early experiments studying the processes occurring inside live cells, we used the formulas discussed in Section 3.2 of the theory. The main long-term objective of the undertaken studies was search of approaches that would permit creation of an optical technique and a device studying the processes in the live cell membranes. The first objective of the experiments was search of cell metabolism parameters. We used an optical setup whose scheme corresponded to the optical scheme discussed in the theory (**Figure 2**). The photo of the setup is shown in **Figure 13**. A semi-conductor laser module with the wavelength λ = 532 nm and 20-mW power as well as the TV camera discussed above in Section 4.2 were used. The setup was placed into a thermostat that maintained the temperature of 36 ± 0.1°C. A transparent cuvette with two glass supporters in a horizontal position in the nutrient solution was fixed on a small table near the mat glass. The first supporter contained a monolayer of cultured cells; the second one was cell-free. Speckle images of the supporters are shown in **Figure 14**. The light lines in the picture show the typical sections selected for determination of value η using

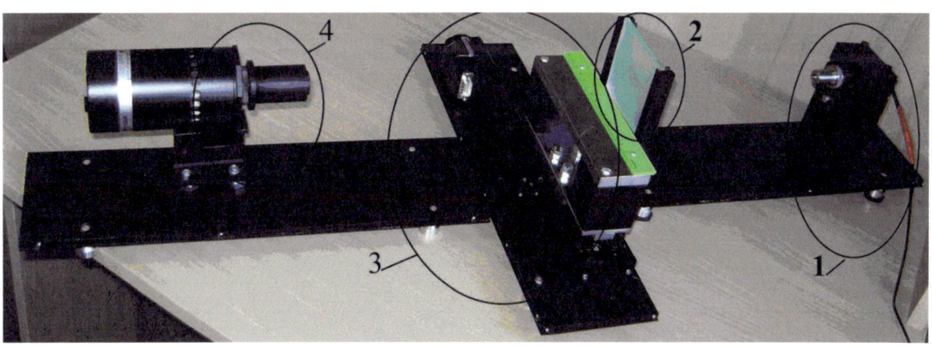

Figure 13.
Photograph of the optical device: 1—laser module with micro-objective, 2—mat glass, 3—lock of the object on the platform of the motorized translator, and 4—camera with lens.

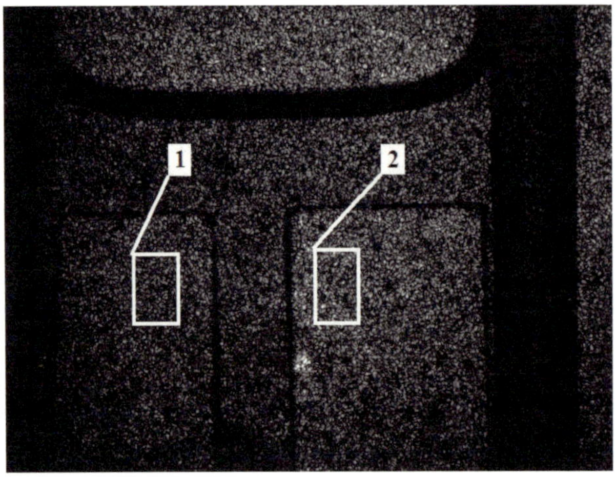

Figure 14.
Speckle images of supports in a transparent cuvette with nutrient solution.

formula (15). The measurements of η were determined using the films recorded during 20–40 s. **Figure 15** presents dependences $\eta(t)$ corresponding to the nutrient solution (top), to the cells in the nutrient solution (middle), and to the herpes simplex-infected cells (bottom). It is seen from the graphs that in all the cases dependence $\eta(t)$ levels off in several seconds. According to the theory, the presence of a constant level speaks for the fact that random radiation intensity variations in the selected image fragment and variations in difference Δu of the optical wave paths in time in the conjugated region are stationary processes. By the measurements of the fixed levels, we determined variances $\sigma^2_{\Delta u}$ of values Δu. In the assumption that random values Δu corresponding to the cells and the nutrient solution are uncorrelated variables, we determined values $\sigma^2_{\Delta u}$ corresponding to the cells.

While processing the films of 20- to 40-s duration recorded for several hours, we detected various types of dependences $\eta(t)$. Along with the graphs similar to those presented in **Figure 15**, we obtained dependences that did not level off as well as graphs that had rather a composite view. All-day graphs of dependences $\sigma_{\Delta u}(t)$ at a 0.5-h pitch were reproducible in about 50% of the cases. Analysis of obtained dependences $\eta(t)$ showed that their complicated character is probably connected to

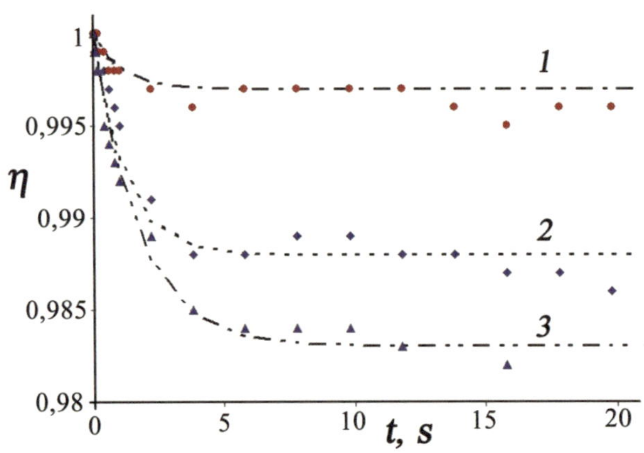

Figure 15.
Dependence $\eta(t)$ corresponding to: 1—nutrient solution, 2—cells in the nutrient solution, and 3—infected cells in the nutrient solution.

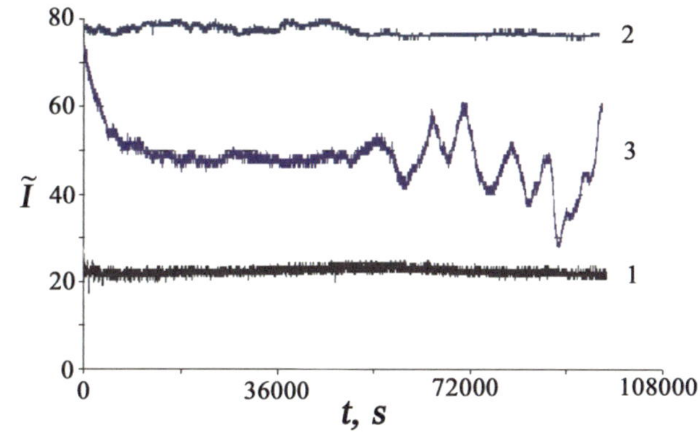

Figure 16.
Dependence of time-averaged intensity $\tilde{I}$ on time for the pixels corresponding to: 1—nutrient solution, 2—cell, and 3—herpes simplex virus in a cell.

the presence of several processes altering the phases of the sounding waves at different rates. In this regard, we used a time-averaged speckle technique discussed in Section 3.3. Application of this technique enabled us to obtain well-reproducible data. **Figure 16** presents typical dependences of the time-averaged digital value of intensity $\tilde{I}$ corresponding to the nutrient solution, to a cell in the nutrient solution and to a herpes simplex virus-infected cell in the solution. The dependences are taken from our paper [38]. Analysis of the dependences belonging to the latter type showed that variation features of value $\tilde{I}$ well corresponds to the stages of virus development in cells. **Figure 17** presents typical dependences $\eta(t)$ corresponding to the nutrient solution, to the cells in the nutrient solution, and to the herpes simplex virus-infected cells in the nutrient medium.

The shown dependences were well-reproducible when a monolayer of cultured cells of various cell lines was infected with herpes virus.

Figure 18 presents joint dependences of $\sigma_{\Delta u}(t)$ and temperature T on time for L-41 cell line [39]. For the linear correlation coefficient of two arrays $\sigma_{\Delta u}$ and T, we obtained a value equal to 0.88. **Figure 14** shows dependence of $\sigma_{\Delta u}$ on T from our early work [35]. L-41 cell line was also used in the experiment.

As the metabolic processes are manifested more distinctly when the temperature rises, the above data presented in [38] were used to substantiate application of value $\sigma_{\Delta u}$ as the parameter that characterizes the activity of cultured cells.

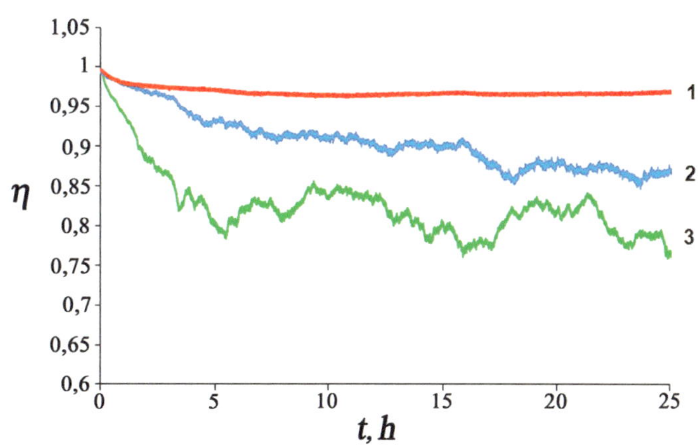

Figure 17.
Typical dependence $\eta(t)$ for Vero line: 1—nutrient solution, 2—virus-free cells, and 3—cells with virus.

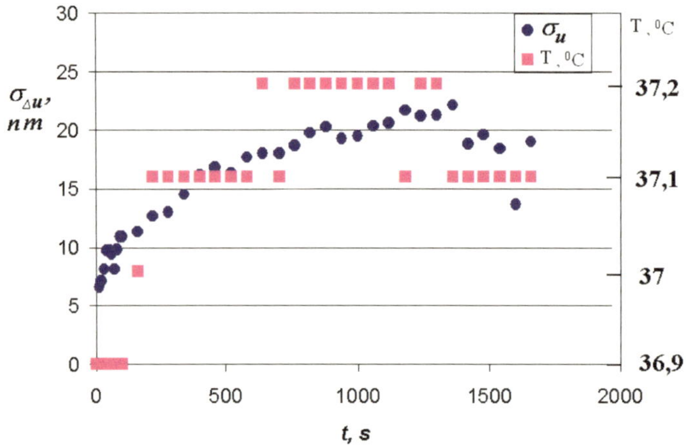

Figure 18.
Joint dependence of $\sigma_{\Delta u}$ and temperature on time at small heating rates.

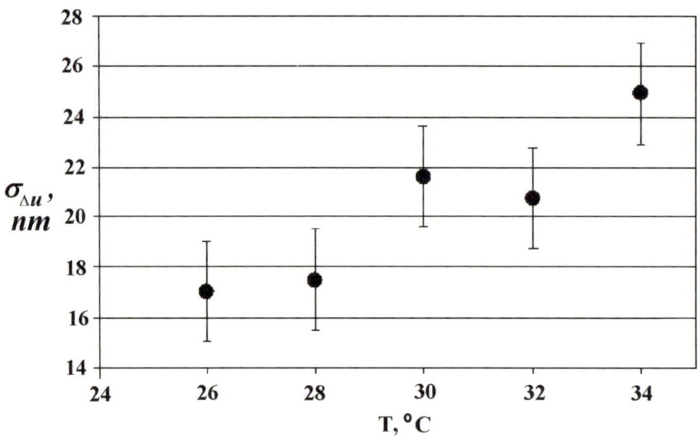

Figure 19.
Dependence of $\sigma_{\Delta u}$ on the temperature.

5. Discussion of results and prospects for further studies

5.1 Studying high-cycle fatigue of metals

The conducted theoretical and experimental research showed that the dynamic variant of speckle interferometry can be used for quantitative evaluation of irreversible displacements and deformations occurring in metals with high-cycle fatigue on the 10-μm order bases. The peculiarity of this evaluation is that application of time-average speckles makes it can be conducted in real time, i.e., without interruption of cyclic deformations (**Figure 19**).

The technique is fairly simple, because it does not require any synchronization of the load value applied to the object and the frame capture moment. It is characterized by high accuracy and sensitivity.

In [32], it was shown that when a 10 × 10-pixel fragment is selected at the variation of η from 0.3 to 0.99, the ratio error of its determination does not exceed 1%. It was shown that a small error in determination of value η by formula (15) is related to the features of radiation intensity $\tilde{I}$ distribution in the speckle pattern. The record error of value η was found to equal the sum of determination errors of values $\tilde{I}$ multiplied by 10^{-2}-order coefficients with different signs. The random meaning of the sign was determined by the difference of $\tilde{I}$ in a pixel and by the mean measurement in the image fragment.

High sensitivity of the speckle technique to deformations is based on the phenomenon of multiple-wave interference with the same initial phases. Speckles emerge as the result of multiple-wave interference with random initial phases. So the emergence of multiple-ray interference is in many ways similar to wave interference in a diffraction grid is not evident. This question was briefly discussed in [34, 40]. Because of the significance of the matter, let us dwell upon it. Let there be a great number of point diffusers on the surface in a region of ΔX diameter with the center at point A, and let value ΔX equal the linear resolution of the lens. For certainty, let the surface be illuminated and observed along the normal. The waves with random initial phases going from the scattering centers form random intensity value $\tilde{I}$ at conjugated point A' of image plane. Now let us imagine that a slight plastic rotation around an axis parallel to oy axis. Let the rotation axis be located at the edge of a region with ΔX diameter. Now let us discuss the scattering centers with the following properties: (1) the waves scattered by these centers have the same amplitudes and initial phases, (2) y-coordinates of the centers are random, (3) along ox axis they are located at the same Δx distance. We suppose that such centers will always be found if their number is large. Let Δu_z be the maximum displacement along the normal in the region, and $M = \Delta X/\Delta x$. In [34], we showed that at point A' the radiation intensity will vary in a quasi-periodic mode proportionally to value $\sin^2(2\pi \times \Delta u_{z0} M/\lambda)/\sin^2(2\pi \times \Delta u_{z0}/\lambda)$, where $\Delta u_{z0} = \Delta u_z/M$. Analysis of the obtained expression showed that in deformation of the materials used in practice, the periodicity corresponding to the denominator is difficult to implement. From the numerator period, we obtain that $\Delta u_{z0} \times M = \Delta u_z = \lambda N/2$, where N is the interference period (period of value $\tilde{I}$). The latter ratio is also implied by formulas (6), (10), (11), if in these formulas we accept that $\sigma^2 = 0$, $\langle x_1 \rangle = 0$, and $k_{11}, k_{22} = 0$. Therefore, the multiple-wave interference that we have discussed is equivalent to interference of two waves spreading from the centers located at the opposite edges of the region with ΔX diameter. Expression $\Delta u_{z0} \times M = \lambda N/2$ may be interpreted in a different way. Having accepted that $N = 1$, we have: $\Delta u_{z0} = (1/2)\lambda/M = (1/2)\lambda_0$. We obtain that variation of value $\tilde{I}$ per period may also be interpreted as the variation result of the interference period of two coherent waves with wave length $\lambda_0 = \lambda/M$. This variation appears in a relative displacement of two scattering centers located at distance Δx by value Δu_{z0}. Let us accept the limiting sensitivity Δu_{z0} of the technique, the 1-nm order is theoretically evaluated above and experimentally confirmed in [41]. Then, formally we obtain that value λ_0 is in the X-ray wave range. This way, if the region with ΔX diameter is so small that the macroscopic deformations in it are homogeneous, then the reflected waves with random initial phases form some random hardly variable speckle brightness value, and the waves with the same initial phases are responsible for quasi-periodic intensity variation on the background of this value.

The conducted experiments showed the possibility in principle to determine the limiting roughness parameter variation ΔR_a, of value $\Delta u = \Delta\vec{u}\left(\vec{l_s} + \vec{l}\right)$ for the reflecting object and the refraction index of a transparent object corresponding to the crack start. By monitoring the variation rate of these parameters on the bases of the 10–100 μm order, one can assess the approximate time to macro-fracture. It is noteworthy that in practice, not always is there an opportunity to illuminate the object under control with laser radiation, and the conventional nondestructive testing techniques are poorly adapted for measurement on such bases. The discussed speckle method could be used as the tool for upgrading of the conventional testing techniques. The technique is convenient for target detection in search of local irreversible deformations and calibration of other techniques.

Fatigue experiments improved understanding of the processes occurring in high-cycle fatigue of materials. For example, the pins that we observed at the notch tip (shown by the arrow in **Figure 9b**) turned out to be pieces of iron carbide. We managed to identify them using Raman scattering [42]. This fact speaks for the significance of heat generation accounting in high-cycle fatigue. The experiments also showed that irreversible processes in the small region close to the notch emerge at the early stages of the fatigue. If inconsistent local deformations emerge at the stage of loading increase, then in arbitrary unloading residual compressive stress must inevitably appear. Thus, the stress and deformation fields at the notch tip will vary considerably with progress of the fatigue. In this context, analysis of not only local plastic, but also of periodically varying elastic deformations is important.

The opportunity for application of the elastic deformation control technique that we have discussed was substantiated in [32]. Deformation $\varepsilon = \Delta u / \Delta X$ of the 10^{-3} order corresponds to values Δu of the 1-nm order and ΔX of the 10-μm order. Plastic deformations 10^{-3} up are more characteristic for most of the conventional constructive materials. If we increase the measurement base by an order, we will proceed into the range of elastic deformations of the 10^{-4} order. It is deformations of this order that emerge in constructions during exploitation. In [43], we conducted a successful pilot model experiment recording values $\tilde{I}$ and η in a specimen under elastic strain. **Figure 20** presents dependences of $\tilde{I}$ and η on the time taken from that paper. The graphs were obtained in periodic deflection of a steel beam. The data correspond to a single oscillation period; the observation and illumination directions selected were not on the surface normal, measurement base ΔX equaled 670 μm. Thus, the approaches that we have discussed are in principle applicable for simultaneous monitoring of both macroscopic elastic deformations on relatively large bases and of local plastic deformations. Development of such techniques can be the subject for further research.

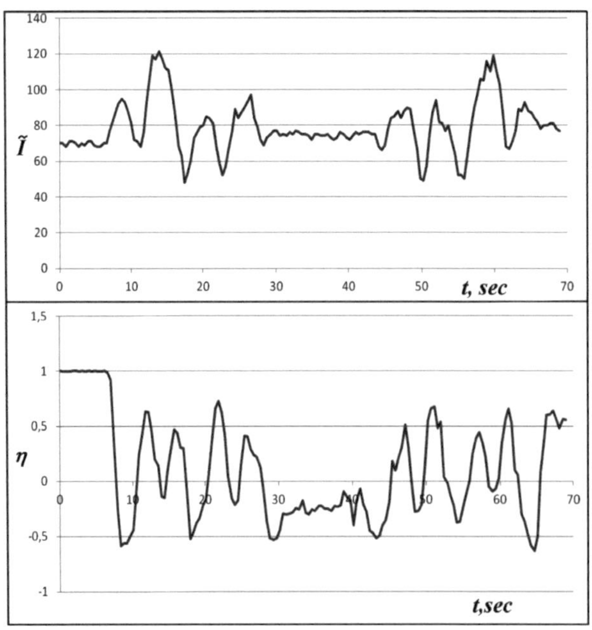

Figure 20.
Dependence of $\tilde{I}$ and η on time in elastic deflection of a steel beam.

The dependence of value η both on the mean value of $x_2 = k\Delta u$ and on its variance k_{22} is the peculiarity of this technique. We managed to obtain numerical values of x_2 and k_{22} for cases $k_{22} = 0$ and $x_2 = 0$, respectively. In [32], it was proposed to obtain the values of $\langle x_2 \rangle$ and k_{22} by simultaneous recording of two speckle images obtained using two lasers. Development of such a technique can be the subject for further research.

In the conducted fatigue experiments, we used an optical system that permits determination mean values, variance, and relaxation time of Δu_z-component of vector $\Delta\vec{u}$. However, the measurements of the above values for other components of vector $\Delta\vec{u}$ are of practical interest. Development of such a technique is the subject of our further research.

5.2 Study of the processes inside live cells

The matters of the accuracy and sensitivity of the technique, the peculiarities of speckle image variations, contribution assessment for values $\langle x_2 \rangle$ and k_{22} to variation of value η discussed in the previous section are fully applied to dynamic speckle interferometry of live cells.

We underpinned and approved a cell activity parameter $\sigma^2_{\Delta u}$, or variance of the difference in the optical path of the cell-sounding waves is such a parameter. However, it is not yet clear what constituents of metabolism affect variation of value $\sigma^2_{\Delta u}$ and to what extent. The study of this problem is the subject of our further research.

Obtaining data by averaging by the cell thickness is a drawback of this technique. Still, the logic of the technique development and the practical needs set the task of determining the mean value, variance, and relaxation time of the medium refraction index in every small section of the cell. The author reported about the ways to solve this problem at two conferences [44, 45]. Theoretical and experimental underpinning of a speckle tomography for the live cell that would permit a solution to this problem is the subject of our further research.

6. Conclusion

This work discussed theoretical and experimental underpinning of an interference technique that permits studying irreversible processes occurring near the surface of reflecting objects and inside thin transparent objects by variation of speckle images. The author's scientific interests lie in the sphere of studying longevity of living and nonliving matter. So, the research targets were specimens made from constructive materials tested for high-cycle fatigue as well as cultured live cells.

The theoretically obtained formulas established the relationship among the parameters characterizing variation of optical wave paths in small sections of an object and the parameters characterizing variation of speckles in the conjugated region. Such parameters for a reflecting object were the mean value, variance, the relaxation time of difference Δu in displacement of scattering centers (points of the surface), time-averaged radiation intensity $\tilde{I}$, and correlation coefficient η for the speckle image fragment taken at the reference and at the current time points. Similar ratios were obtained for a transparent object sounded by multiple waves with random initial phases. The difference consisted of the fact that value Δu was the difference in the optical path of the waves sounding the object.

For the variant, when variation of Δu in time is a random stationary process, we obtained a relation between the normalized spectrum of value Δu and the normalized fluctuation spectrum of $\tilde{I}$.

In experiments set up for testing of the theory, speckle dynamics was created by displacement of a rough transparent plate, by variation of shape, roughness, and the refraction index of metal and Plexiglas specimens in their high-cycle fatigue. Good quantitative coincidence of the theory and the experiment was shown.

It was shown that by way of speckle time-averaging the technique permits real-time determination of the 1-nm-order measurements of Δu on the 10-μm order on bases ΔX, which corresponds to local plastic deformation $\varepsilon = \Delta u/\Delta X$ of the 10^{-3} order.

It was demonstrated that with increasing base ΔX by an order, the technique allows to monitor the variation of elastic macroscopic deformations. So, the possibility, in principle, to use the same technique for monitoring both the deformation field of a part and accumulation of local plastic deformations in an object under periodic strain appears.

It was also shown that using this technique, one can determine the limiting variations of roughness parameter R_a, value ε and refraction index n that correspond to the crack start. Therefore, knowing the limiting measurements of these parameters and monitoring their variation rate while exploiting the part, one could in principle assess the time to its macroscopic fracture. Development of such a technique can be the subject of further research.

When the technique was applied for studying live cells, it was shown that the variance of value Δu can be used as a live cell activity parameter. It was also shown that dependences $\tilde{I}(t)$ and $\eta(t)$ for virus-infected and virus-free cells differ considerably. Modification of this technique for determination of the average measurement, variance, and relaxation time of the refraction index in small areas of the live cell can be the line of further research.

Acknowledgements

The author would like to thank the young colleagues and students I. Kamantsev, N. Drukarenko, Yu. Mikhailova, and K. Myznov for their help in conducting experiments.

Author details

Alexander Vladimirov
Ural Federal University, Yekaterinburg Institute of Virus Infections, Yekaterinburg, Russia

*Address all correspondence to: vap52@bk.ru

References

[1] Anisimov VI, Kozel SM, Lokshin GR. Space-time statistical properties of the coherent radiation scattered by a moving diffuse reflector. Optics and Spectroscopy. 1969;**27**(3):483-491

[2] Yoshimura T. Statistical properties of dynamic speckles. JOSA. 1986;**A3**(7): 1032-1054

[3] Yamaguchi I. Speckle displacement and decorrelation in the diffraction and image fields for small object deformation. Optica Acta. 1991;**28**: 1359-1376

[4] Aleksandrov EB, Bonch-Bruevich AM. Investigation of surface strains by the hologram technique. Soviet Physics-Technical Papers. 1967;**12**: 258-265

[5] Leendertz JA. Interferometric displacement measurement on scattering surfaces utilizing speckle effect. Journal of Physics E: Scientific Instruments. 1970;**3**:214-218

[6] Vladimirov AP, Mikushin VI. Interferometric determination of vector components of relative displacements: Theory and experiment. Proceedings of SPIE. 1999;**3726**:38-43

[7] Fomin NA. Speckle Photography for Fluid Mechanics Measurements. Berlin: Springer-Verlag; 1998. 219p

[8] Romanov AN. Razrushenie pri Malotsiklovom Nagruzhenii [Fracture in Low-Cycle Loading]. Moscow: Nauka; 1988. 282p; (rus)

[9] Ellyin F. Fatigue damage, crack growth life prediction. London: Chapman and Hall; 1997. 483p

[10] Brinckmann S. On the role of dislocations in fatigue crack initiation [Dissertation]. Groningen, The Netherlands: University of Groningen; 2005

[11] Novikov II, Yermishin VA. Fizicheskaia Mehanika Realnyh Materialov [Physical Mechanics of Real Materials]. Moscow: Nauka; 2004. 328p; (rus)

[12] Lasar J, Hola M, Cip O. Differential interferometry for real-time measurement in high cycle fatigue metal testing. Book of abstracts. In: Conference on PhotoMechanics 2015. 25–27 May 2015. Delft, Netherlands. pp. 64-65

[13] Gough HJ. Fatigue of Metals. London: Benn; 1926

[14] Ustalost metallov. Sbornik statei pod red, G.V. Uzhika [Metal fatigue. Collected papers ed. By G.V. Uzhik]. Moscow: Izd-vo inostr.lit. – Moscow, 1961. 378 p. (rus)

[15] Ivanova VS, Terentiev VF. Priroda Ustalosti Materialov [The Nature of Metal Fatigue]. Moscow: Metallurgija; 1975. 456p; (rus)

[16] Taylor D. The Theory of Critical Distances. Amsterdam: Elsevier; 2007. 307p

[17] Manson SS. Fatigue: A complex subject—Some simple approximations. The William M. Murray Lecture. Experimental Mechanics. 1965;5(7): 193-226

[18] Schijve J. Fatigue of structures and materials in the 20th century and the state of art. International Journal of Fatigue. 2003;**25**(8):679-702

[19] Terentiev VF. Ustalost Metallicheskih Materialov [Fatigue of Metallic Materials]. Moscow: Nauka; 2002. 248p; (rus)

[20] Gilanyi A, Morishita K, Sukegawa T, Uesaka M, Miya K. Magnetic non-destructive evaluation of fatigue damage of ferromagnetic steels for nuclear fusion energy systems. Fusion Engineering and Design. 1998;**42**:485-491

[21] Gorkunov ES, Savray RA, Makarov AV, Zadvorkin SM. Magnetic techniques for estimating elastic and plastic strains in steels under cyclic loading. Diagnostics, Resource and Mechanics of Materials and Structures. 2015;**2**:6-15

[22] Tupikin DA. Thermoelectric method of fatigue phenomena control. Control Diagnostika. 2003;**11**:53-61. (rus)

[23] Ignatovich VN, Shmarov SS, Yutskevich SR. Features of formation of deformation relief on the surface of the alloy D16AT fatigue. Aviatsionno-Kosmicheslaiia Tehnika i Tehnologiia. 2009;**67**(10):132-135. (rus)

[24] Yermishkin VA, Murat DP, Podbelskiy VV. Application of photometric analysis of structural images to assess fatigue resistance. Avtomatizatsija i Sovremennije Tehnologii. 2008;**2**:11-21. (rus)

[25] Plehov OA, Panteleiev IA, Leontiev VA. Features of heat release and generation of acoustic emission signals during cyclic deformation of Armco-iron. Fizicheskaija Mezomehanika. 2009;**12**(5):37-43. (rus)

[26] Marom E. Holographic Correlation. In: Erf PK, editor. Holographic Nondestructive Testing. New York: Academic Press; 1974. p. 149-180

[27] Marom E, Muller RK. Optical correlation for impending fatigue failure detection. International Journal of Nondestructive Testing. 1971;**3**:171-187

[28] Kozubenko VP, Potichenko VA, Borodin YS. Study of metal fatigue by speckle-correlation method. Problemy Prochnosti. 1989;**7**:103-107. (rus)

[29] Vladimirov AP. Dinamika speklov v ploskosti izobrazhenija plasticheski deformiruiemogo ob'ekta. Diss... cand. fiz.-mat. nauk [Speckle dynamics in the image plane of the deformed object] [Dissert. Cand. Sci]. Sverdlovsk; 1986. 106p; (rus)

[30] Vladimirov AP. Dinamicheskaia spekl-interferometrija deformiruiemyh ob'iektov. Diss... dokt. tehn. nauk [Dynamic speckle interferometry of deformed objects] [Dissert. Doc. Eng. Sci.] Yekaterinburg; 2002. 393p; (rus)

[31] Vladimirov AP. Dynamic speckle interferometry of the microscopic processes. Proceedings of SPIE. 2012; **8413**:841305; 1-6

[32] Vladimirov AP. Speckle metrology of dynamic macro- and micro-processes in deformable media. Optical Engineering. 2016;**55**(12):121727; 1-10. DOI: 10.1117/1.OE.55.12.121727

[33] Vladimirov AP, Druzhinin AV, Malygin AS, Mikitas' KN. Theory and calibration of speckle dynamics of phase object. SPIE Proceedings. 2012;**8337**: 8337OC; 1-15

[34] Vladimirov AP. Dynamic speckle – Interferometry of micro-displacements. AIP Conference Proceedings. 2012;**1457**: 459-468

[35] Malygin AS, Bebenina NV, Vladimirov AP, Mikitas' KN, Bakharev AA. A speckle_interferometric device for studying the cell biological activity. Instruments and Experimental Techniques. 2012;**55**(3):415-418

[36] Vladimirov AP. Dynamic speckle interferometry of microscopic processes in thin biological objects. Radiophysics and Quantum Electronics. 2015;**57**(8): 564-576

[37] Vladimirov AP. Dynamic speckle interferometry of high-cycle material fatigue: Theory and some experiments. AIP Conf. Proc. 2016;**1740**:040004. DOI: 10-1063/14962663

[38] Vladimirov AP, Bakharev AA. Dynamic speckle interferometry of thin biological objects: Theory, experiments and practical perspectives. In: Banishev AA, Bhowmick M, Wang J, editors. Optical Interferometry. Rijeka, Croatia: InTech; 2017. pp. 103-141. Print ISBN: 978-953-51-2955-4; Online ISBN: 978-953-51-2956-1

[39] Mikhailova YA, Vladimirov AP, Bakharev AA, Sergeev AG, Novoselova IA, Yakin DI. Study of cell culture reaction to temperature change by dynamic speckle interferometry [Electronic resource]. Rossijskij Zhurnal Biomehaniki. 2017;**21**(1):64-73

[40] Vladimirov AP, Kamantsev IS, Veselova VE, Gorkunov ES, Gladkovskii SV. Use of dynamic speckle interferometry for contactless diagnostics of fatigue crack initiation and determining its growth rate. Technical Physics. 2016;**61**(4): 563-568

[41] Vladimirov AP, Kamantsev IS, Ishchenko AV, Gorkunov ES, Gladkovskii SV, Zadvorkin SM. The study of the process of nucleation of fatigue cracks by changing the surface topography of the sample and its speckle images. Deformatsija i Razrushenije Materialov. 2015;**1**:21-26

[42] Tihonova IN, Men'shchikova AA, Vladimirov AP, Ponosov YS, Kamantsev IS, Drukarenko NA, Ishchenko AV. Kompleksnoie izuchenie mnogotsiklovoi ustalosti stali metodami dinamiki speklov, opticheskoi profilometrii, konfokal'noi, skaniruiiushei, magnitnoi I ramanovskoi mikroscopii, XXX Mezhdunarodnaia shkola-simposium po golographii, kogerentnoi optike I fotonike: Materialy shkoly-simposiuma. Pod red. kand. fiz.-mat. nauk I.V. Alekseienko [In-depth study of high-cycle fatigue of steel using speckle dynamics, optical profilometry, confocal, scanning, magnetic and Raman microscopy. In: XXX International symposium school in holography, coherent optics and photonics: Proceedings of symposium school. Ed. by Cand. Sci. I.V. Alekseienko]. Kaliningrad: BFU; 2017. pp. 122-123

[43] Vladimirov AP, Korovin BB, Chervoniuk VV. K razrabotke sredstv kontrolia vibratsiy rabochih lopatok GTD s ispolzovaniiem metoda dinamicheskogo speckl-interferometricheskogo opredeleniia deformatsii. Perspectivy razvitiia aviatsionnyh kompleksov gosudarstvennoi aviatsii i ih silovyh ustanovok. Sbornik nauchnyh statei po materialam V Mezhdunarodnoi nauchno-practicheskoi konferentsii "Zhukovskiie akademicheskiie chteniia" [On developing facilities for monitoring the vibrations of GTE blades using speckle interferometry for strain determination/Development prospects of state aircraft systems and their power units. Collected papers based on Proceedings of V International Applied Science Conference "Zhukovsky Academic readings"]; 22–23 Nov. 2017. Voronezh; 2018. pp. 79-84

[44] Vladimirov AP. Chetyrehmernaiia spekl-tomografiia zhivyh I tehnicheskih tonkih ob'iektov: Teoreticheskaia baza, eksperimenty, dostizheniia i problemy. XXX Mezhdunarodnaia shkola-simposium po golographii, kogerentnoi optike I fotonike: Materialy shkoly-simposiuma . Pod red. kand. fiz.-mat. nauk I.V. Alekseienko [Four-dimensional speckle tomography of live and technical thin objects: Theoretical underpinning, experiments, advances and problems. In: XXX International symposium school in holography, coherent optics and photonics: Proceedings of symposium school. Ed.

by Cand. Sci. I.V. Alekseienko]. Kaliningrad: BFU; 2017. pp. 48-49

[45] Vladimirov AP. Four-dimensional speckle-interferometric microscopy of thin biological objects. In: Book of Abstracts of 5th International Congress on Microscopy & Spectroscopy (INTERM 2018). Oludeniz, Turkey. April 24–30. 2018; p. 36